Millimeter-Wave (mmWave) Communications

Millimeter-Wave (mmWave) Communications

Special Issue Editor
Manuel García Sanchez

MDPI • Basel • Beijing • Wuhan • Barcelona • Belgrade

Special Issue Editor
Manuel García Sanchez
Signal Theory and
Communications Department,
University of Vigo
Spain

Editorial Office
MDPI
St. Alban-Anlage 66
4052 Basel, Switzerland

This is a reprint of articles from the Special Issue published online in the open access journal *Electronics* (ISSN 2079-9292) from 2019 to 2020 (available at: https://www.mdpi.com/journal/electronics/special_issues/mmwave_commun).

For citation purposes, cite each article independently as indicated on the article page online and as indicated below:

LastName, A.A.; LastName, B.B.; LastName, C.C. Article Title. *Journal Name* **Year**, *Article Number*, Page Range.

ISBN 978-3-03928-430-6 (Pbk)
ISBN 978-3-03928-431-3 (PDF)

Cover image courtesy of Manuel García Sánchez.

© 2020 by the authors. Articles in this book are Open Access and distributed under the Creative Commons Attribution (CC BY) license, which allows users to download, copy and build upon published articles, as long as the author and publisher are properly credited, which ensures maximum dissemination and a wider impact of our publications.

The book as a whole is distributed by MDPI under the terms and conditions of the Creative Commons license CC BY-NC-ND.

Contents

About the Special Issue Editor vii

Preface to "Millimeter-Wave (mmWave) Communications" ix

Manuel García Sánchez
Millimeter-Wave Communications
Reprinted from: *Electronics* **2020**, *9*, 251, doi:10.3390/electronics9020251 1

Lorenzo Rubio, Rafael P. Torres, Vicent M. Rodrigo Peñarrocha, Jesús R. Pérez, Herman Fernández, Jose-Maria Molina-Garcia-Pardo and Juan Reig
Contribution to the Channel Path Loss and Time-Dispersion Characterization in an Office Environment at 26 GHz
Reprinted from: *Electronics* **2019**, *8*, 1261, doi:10.3390/electronics8111261 4

Miguel Riobó, Rob Hofman, Iñigo Cuiñas, Manuel García Sánchez and Jo Verhaevert
Wideband Performance Comparison between the 40 GHz and 60 GHz Frequency Bands for Indoor Radio Channels
Reprinted from: *Electronics* **2019**, *8*, 1234, doi:10.3390/electronics8111234 18

Benxiao Cai, Lingling Sun and Yuchao Lei
3D Printing Using a 60 GHz Millimeter Wave Segmented Parabolic Reflective Curved Antenna
Reprinted from: *Electronics* **2019**, *8*, 203, doi:10.3390/electronics8020203 41

Andrea Massaccesi, Gianluca Dassano and Paola Pirinoli
Beam Scanning Capabilities of a 3D-Printed Perforated Dielectric Transmitarray
Reprinted from: *Electronics* **2019**, *8*, 379, doi:10.3390/electronics8040379 50

Sara Salem Hesari and Jens Bornemann
Design of a SIW Variable Phase Shifter for Beam Steering Antenna Systems
Reprinted from: *Electronics* **2019**, *8*, 1013, doi:10.3390/electronics8091013 63

Heng Zhu, Wei Chen, Jianhua Huang, Zhiyu Wang and Faxin Yu
A High-Efficiency K-band MMIC Linear Amplifier Using Diode Compensation
Reprinted from: *Electronics* **2019**, *8*, 487, doi:10.3390/electronics8050487 77

Jihoon Doo, Woojin Park, Wonseok Choe and Jinho Jeong
Design of Broadband W-Band Waveguide Package and Application to Low Noise Amplifier Module
Reprinted from: *Electronics* **2019**, *8*, 523, doi:10.3390/electronics8050523 89

Xiaofan Yang, Xiaoming Liu, Shuo Yu, Lu Gan, Jun Zhou and Yonghu Zeng
Permittivity of Undoped Silicon in the Millimeter Wave Range
Reprinted from: *Electronics* **2019**, *8*, 886, doi:10.3390/electronics8080886 98

Daniel Castanheira, Sara Teodoro, Ricardo Simões, Adão Silva and Atilio Gameiro
Multi-User Linear Equalizer and Precoder Scheme for Hybrid Sub-Connected Wideband Systems
Reprinted from: *Electronics* **2019**, *8*, 436, doi:10.3390/electronics8040436 110

Adel Aldalbahi
Multi-Backup Beams for Instantaneous Link Recovery in mmWave Communications
Reprinted from: *Electronics* **2019**, *8*, 1145, doi:10.3390/electronics8101145 . . . . **126**

Enass Hriba and Matthew C. Valenti
Correlated Blocking in mmWave Cellular Networks: Macrodiversity, Outage, and Interference
Reprinted from: *Electronics* **2019**, *8*, 1187, doi:10.3390/electronics8101187 . . . . **137**

Luis Duarte, Rodolfo Gomes, Carlos Ribeiro and Rafael F. S. Caldeirinha
A Software-Defined Radio for Future Wireless Communication Systems at 60 GHz
Reprinted from: *Electronics* **2019**, *8*, 1490, doi:10.3390/electronics8121490 . . . . **154**

About the Special Issue Editor

Manuel García Sánchez received Telecommunication Engineering degree from the Universidade de Santiago de Compostela, Santiago de Compostela, Spain, in 1990, and a Ph.D. in Telecommunication Engineering from the Universidade de Vigo, Spain, in 1996. He is currently a professor at the Department of Signal Theory and Communications, Universidade de Vigo, Spain. He was head of the department from 2004 to 2010. His research interests include radio systems, indoor and outdoor radio channels, channel sounding and modeling for narrowband and wideband applications, interference detection and analysis, design of impairment mitigation techniques, and radio systems design.

Editorial

Millimeter-Wave Communications

Manuel García Sánchez

atlanTTic Research Center, Signal Theory and Communications Department, University of Vigo, 36310 Vigo, Spain; manuel.garciasanchez@uvigo.es

Received: 24 January 2020; Accepted: 29 January 2020; Published: 3 February 2020

1. Introduction

For the last few decades, the millimeter wave (mmWave) frequency band (30–300 GHz) has been seen as a serious candidate to host very high data rate communications. First used for high capacity radio links, then for broadband indoor wireless networks, the interest in this frequency band has boosted, as it is proposed to accommodate future 5G mobile communication systems. The large bandwidth available in this frequency band will enable a number of new use cases for 5G. In addition, due to the large propagation attenuation, this frequency band may present some additional advantages regarding frequency reuse and communication security. On the other hand, however, a number of issues have to be addressed to make 5G mmWave communications viable: radio channel measurement, modeling, and estimation; antenna design and antenna measurement; beamforming and energy efficiency; commercial hardware design and development; multiple-input multiple-output (MIMO) and massive MIMO (m-MIMO) techniques; multi-cell cooperation; network planning and interference; system performance assessment and optimization; and finally, the study of new case uses and applications.

2. Contributions in This Special Issue

Each of the twelve papers collected in this Special Issue contributes to a solution to one or more of the challenges described in the introduction. Regarding radio wave propagation, Rubio et al. [1] provide an experimental characterization of the path loss and time-dispersion of an in office radio channel at 26 GHz, while in a study by Riobó and colleagues [2], wideband results at 40 GHz and 60 GHz frequency bands are also provided for indoor environments.

Two other papers [3,4] deal with the design and assessment of different kinds of antennas manufactured using three-dimensional (3D) printing. In one case, a 60 GHz segmented parabolic reflective curved antenna, with a gain of 20 dBi at 64 GHz, is presented by Cai, Sun, and Lei [3], while Massaccesi, Dassano, and Pirinoli [4] design a perforated dielectric transmitarray and analyze its beam scanning capabilities. Also related to beam steering antennas, Salem Hesari and Bornemann [5] describe the design, fabrication, and assessment of a substrate integrated waveguide variable phase shifter that may steer the radiation pattern of the antenna by ±25°.

Another challenge for mmWave communication system is the design of amplifiers. Two different kinds of mmWave amplifiers are presented in two contributions. The design of a high-efficiency K-band MMIC linear amplifier using diode compensation is presented by Zhu and co-workers [6] together with its measured performance, while in a study by Doo and colleagues [7] a broadband mmWave waveguide package, which covers the entire W-band (75–110 GHz), is presented and applied to build a low noise amplifier module. This module measures gains greater than 14.9 dB from 75 GHz to 105 GHz (12.9 dB at the entire W-band) and noise figures less than 4.4 dB from 93.5 GHz to 94.5 GHz.

For a proper design of the electronic systems at mmWave frequencies a good empirical characterization of the dielectric properties of the substrate material is of capital importance. In a study by Yang et al. [8] a description of the dielectric measurement of undoped silicon in the E-band (60–90 GHz) using a free-space quasi-optical system is provided.

Precoding is other key technology that will enable 5G mmWave communications. Castanheira and co-workers [9] present a sub-connected hybrid analog/digital multi-user linear equalizer combined with an analog precoder to efficiently remove the multi-user interference.

As propagation environment imposes several restrictions to radio wave propagation at mmWave frequencies and strong link blockage may occur, any technique to facilitate link recovery constitutes a significant contribution. A multibeam technique to speed up link recovery is presented by Aldalbahi [10]. Hriba and Valenti [11] discuss another way to mitigate link blocking by using macrodiversity techniques, however the performance of macrodiversity can be reduced if correlated blocking occurs in links to different base stations.

Finally, Duarte and colleagues [12] present a complete end-to-end 5G mmWave testbed fully reconfigurable based on a FPGA architecture.

3. Future Trends

The development of millimeter-wave communication systems has just started. Despite the recent developments to cope with the multiple challenges that researchers should solve, there is still a lot of work to be done. During the next years, we hope to assist in the exponential growth of contributions to this field: mm-Wave communications will lead us to full development of 5G case studies and beyond.

Acknowledgments: I would like to thank, in the first place, all the authors who decided to send they research contributions to this Special Issue. I also would like to thank all the reviewers for their comments that helped to improve the papers in this issue. Finally, I would like to thank MDPI editorial board and staff for the opportunity to guest-edit this Special Issue.

Conflicts of Interest: The author declares no conflicts of interest.

References

1. Rubio, L.; Torres, R.; Rodrigo Peñarrocha, V.; Pérez, J.; Fernández, H.; Molina-Garcia-Pardo, J.M.; Reig, J. Contribution to the Channel Path Loss and Time-Dispersion Characterization in an Office Environment at 26 GHz. *Electronics* **2019**, *8*, 1261. [CrossRef]
2. Riobó, M.; Hofman, R.; Cuiñas, I.; García Sánchez, M.; Verhaevert, J. Wideband Performance Comparison between the 40 GHz and 60 GHz Frequency Bands for Indoor Radio Channels. *Electronics* **2019**, *8*, 1234. [CrossRef]
3. Cai, B.; Sun, L.; Lei, Y. 3D Printing Using a 60 GHz Millimeter Wave Segmented Parabolic Reflective Curved Antenna. *Electronics* **2019**, *8*, 203. [CrossRef]
4. Massaccesi, A.; Dassano, G.; Pirinoli, P. Beam Scanning Capabilities of a 3D-Printed Perforated Dielectric Transmitarray. *Electronics* **2019**, *8*, 379. [CrossRef]
5. Salem Hesari, S.; Bornemann, J. Design of a SIW Variable Phase Shifter for Beam Steering Antenna Systems. *Electronics* **2019**, *8*, 1013. [CrossRef]
6. Zhu, H.; Chen, W.I.; Huang, J.; Wang, Z.; Yu, F. A High-Efficiency K-band MMIC Linear Amplifier Using Diode Compensation. *Electronics* **2019**, *8*, 487. [CrossRef]
7. Doo, J.; Park, W.; Choe, W.; Jeong, J. Design of Broadband W-Band Waveguide Package and Application to Low Noise Amplifier Module. *Electronics* **2019**, *8*, 523. [CrossRef]
8. Yang, X.; Liu, X.; Yu, S.; Gan, L.; Zhou, J.; Zeng, Y. Permittivity of Undoped Silicon in the Millimeter Wave Range. *Electronics* **2019**, *8*, 886. [CrossRef]
9. Castanheira, D.; Teodoro, S.; Simões, R.; Silva, A.; Gameiro, A. Multi-User Linear Equalizer and Precoder Scheme for Hybrid Sub-Connected Wideband Systems. *Electronics* **2019**, *8*, 436. [CrossRef]
10. Aldalbahi, A. Multi-Backup Beams for Instantaneous Link Recovery in mmWave Communications. *Electronics* **2019**, *8*, 1145. [CrossRef]

11. Hriba, E.; Valenti, M.C. Correlated Blocking in mmWave Cellular Networks: Macrodiversity, Outage, and Interference. *Electronics* **2019**, *8*, 1187. [CrossRef]
12. Duarte, L.; Gomes, R.; Ribeiro, C.; Caldeirinha, R. A Software-Defined Radio for Future Wireless Communication Systems at 60 GHz. *Electronics* **2019**, *8*, 1490. [CrossRef]

© 2020 by the author. Licensee MDPI, Basel, Switzerland. This article is an open access article distributed under the terms and conditions of the Creative Commons Attribution (CC BY) license (http://creativecommons.org/licenses/by/4.0/).

Article

Contribution to the Channel Path Loss and Time-Dispersion Characterization in an Office Environment at 26 GHz

Lorenzo Rubio [1,*], **Rafael P. Torres** [2], **Vicent M. Rodrigo Peñarrocha** [1], **Jesús R. Pérez** [2], **Herman Fernández** [3], **Jose-Maria Molina-Garcia-Pardo** [4] and **Juan Reig** [1]

1. iTEAM Research Institute, Universitat Politècnica de València, 46022 Valencia, Spain; vrodrigo@dcom.upv.es (V.M.R.P.); jreigp@dcom.upv.es (J.R.)
2. Departamento de Ingeniería de Comunicaciones, Universidad de Cantabria, 39005 Santander, Spain; rafael.torres@unican.es (R.P.T.); jesusramon.perez@unican.es (J.R.P.)
3. Escuela de Ingeniería Electrónica, Universidad Pedagógica y Tecnológica de Colombia, Sogamoso 152211, Colombia; herman.fernandez@uptc.edu.co
4. Departamento de Tecnologías de la Información y las Comunicaciones, Universidad Politécnica de Cartagena, Cartagena, 30202 Murcia, Spain; josemaria.molina@upct.es
* Correspondence: lrubio@dcom.upv.es; Tel.: +34-963879739

Received: 7 October 2019; Accepted: 28 October 2019; Published: 1 November 2019

Abstract: In this paper, path loss and time-dispersion results of the propagation channel in a typical office environment are reported. The results were derived from a channel measurement campaign carried out at 26 GHz in line-of-sight (LOS) and obstructed-LOS (OLOS) conditions. The parameters of both the floating-intercept (FI) and close-in (CI) free space reference distance path loss models were derived using the minimum-mean-squared-error (MMSE). The time-dispersion characteristics of the propagation channel were analyzed through the root-mean-squared (rms) delay-spread and the coherence bandwidth. The results reported here provide better knowledge of the propagation channel features and can be also used to design and evaluate the performance of the next fifth-generation (5G) networks in indoor office environments at the potential 26 GHz frequency band.

Keywords: 5G; mmWave; path loss; time-dispersion; delay-spread; coherence bandwidth; channel measurements

---

1. Introduction

Some of the main objectives proposed in the deployment of the future fifth-generation (5G) systems are the increase in the data rate and capacity, greater than 100 Mbps, with peak data rates up to 10 Gbps, ultra-reliable and low-latency communications, and communications in high user density scenarios [1,2]. The first 5G deployments, at least at the European level, will be carried out in the harmonized frequency bands below 1 GHz, in particular the 700 MHz band, corresponding to the second digital dividend, together with the 3.4–3.8 GHz frequency band [3]. However, the high transmission rates expected in 5G can only be achieved using the spectrum at frequencies above 24 GHz, where it is possible to use bandwidths of the order of hundreds of megahertz [2]. At the last World Radiocommunication Conference (WRC) of the International Telecommunication Union (ITU), held in 2015 (WRC-15), the potential bands to locate future 5G systems, on a primary basis, above 24 GHz are: 24.25–27.5 GHz, 31.8–33.4 GHz, and 37–40.5 GHz [4]. Although the final decision will be conditioned in part by the industry, the potential for global harmonization, and research works, there is some consensus to deploy the 5G systems in the 26 GHz frequency band. In fact, the Radio Spectrum Policy Group (RSPG) has recommended the 24.25–27.5 GHz band for the 5G deployments in Europe.

Many measurement campaigns in both indoor and outdoor environments have been conducted at some typical millimeter wave (mmWave) bands (although the mmWave band strictly corresponds to frequencies between 30 and 300 GHz, it is common in the literature to also consider frequencies above 10 GHz), in particular at 11, 15, 28, 38, 60, and 73 GHz [5–9]. Nevertheless, little attention has been devoted to the 26 GHz band. Although the propagation characteristics measured at 28 GHz could be extrapolated to the 26 GHz frequency band, specific channel measurements are necessary for better knowledge of the propagation channel. In this sense, a contribution to the path loss and time-dispersion characterization, in terms of the delay-spread and the coherence bandwidth, is performed in this paper. The study is based on a channel measurement campaign at 26 GHz carried out in an indoor office environment. The measurements were collected under line-of-sight (LOS) and obstructed-LOS (OLOS) conditions with a channel sounder implemented in the frequency domain using a vector network analyzer (VNA) and a broadband radio over fiber (RoF) link to increase the dynamic range in the measurement and allowing us to use omnidirectional antennas at the transmitter (Tx) and the receiver (Rx).

The remainder of the paper is organized as follows. Section 2 describes the propagation environment, measurement setup, and procedure. In Section 3, path loss and time-dispersion results are presented and discussed. Finally, conclusions are drawn in Section 4.

2. Channel Measurements

2.1. Propagation Environment

The channel measurements were carried out in an office environment, characterized by the presence of desks, with PC monitors, chairs, and some steel storage cabinets, among other typical objects in these environments. The office was in a modern building construction with large exterior glass windows, where the ceiling and the floor were built of reinforced concrete over steel plates with wood and plasterboard paneled walls. Figure 1 shows a panoramic view of the office. The propagation environment consisted of a 9.68 m by 6.93 m room with a height of 2.63 m.

Figure 1. Panoramic view of the propagation environment.

2.2. Measurement Setup and Procedure

A channel sounder was implemented in the frequency domain to measure the complex channel transfer function (CTF), denoted as $H(f)$. The channel sounder was based on the Keysight N5227A VNA. The QOM-SL-0.8-40-K-SG-L ultra-wideband antennas, developed by Steatite Ltd company, were used at the Tx and Rx sides. These antennas operate from 800 MHz to 40 GHz, have an omnidirectional radiation pattern in the azimuth plane (horizontal plane), and linear polarization. Figure 2 shows the three-dimensional (3D) radiation pattern of the antennas measured in our anechoic chamber. The 3 dB beamwidth of the antennas in the elevation plane, known as half power beamwidth (HPBW), was in the order of 35° at 26 GHz.

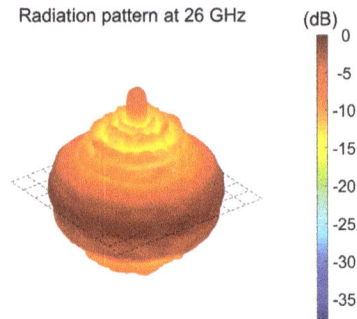

Figure 2. 3D radiation pattern of the antenna at 26 GHz.

The Tx antenna was connected to the VNA through an amplified broadband RoF link, the Optica OTS-2 model developed by Emcore (from 1 to 40 GHz, with 35 dB of gain). This RoF link avoided the high losses of cables at mmWave frequencies, thus increasing the dynamic range in the measurement and allowing us to use omnidirectional antennas due to the fact that significant features of the propagation channel, such as time-dispersion, could be affected by the use of directional antennas [10].

The Rx antenna was located in a XY positioning system, implementing a 12 × 12 uniform rectangular array (URA). The inter-element separation of the URA was 3.04 mm. This separation was about $\lambda/4$ at 26 GHz, covering the Rx antenna with a local area around $(11/4)^2 \lambda^2$ in order to take into account small-scale fading effects. Both the VNA and the XY positioning system were controlled by a personal computer. The $S_{21}(f)$ scattering parameter, equivalent to $H(f)$ [11], was measured from 25 to 27 GHz, i.e., a channel bandwidth (SPAN in the VNA) of 2 GHz was employed with 26 GHz as a central frequency. Notice that a VNA measures the scattering parameters of a device under test (DUT). In this case, the DUT was the wireless channel, where the $S_{21}(f)$ scattering parameter was the CTF at the frequency that was used to excite the channel. By having the excitation signal sweep through the frequency band of interest, i.e., the SPAN, a sampled version of the CTF was measured. The radiofrequency signal level at the VNA was −17 dBm to not saturate the amplifier at the input of the electro-optical converter in the RoF link. Before the measurements, the channel sounder was calibrated carefully. A response calibration process was performed, moving the time reference points from the VNA port to the calibration points. Thus, the measured CTF took into account the joint response of the propagation channel and the Tx and Rx antennas, also known in the literature as the radio channel [12]. A schematic diagram of the channel sounder is shown in Figure 3.

A total of $N_f = 1091$ frequency points was measured over the 2 GHz bandwidth. Thus, the frequency resolution was about $\Delta f \approx 1.83$ MHz (2 GHz/N_f), which corresponded to a maximum unambiguous excess delay estimated as $1/\Delta f$ of 546 ns. This maximum unambiguous excess delay was equivalent to a maximum observable distance calculated as $c_0/\Delta f$, with c_0 the speed of light, of about 164 m. Notice that the maximum observable distance was larger than the office dimensions, ensuring that all multipath contributions were captured. The bandwidth of the intermediate frequency (IF) filter at the VNA, denoted by B_{IF}, was set to 100 Hz. This value of B_{IF} was a trade-off between acquisition time and dynamic range in the measurement. Thus, low values of B_{IF} reduced the noise floor, increasing the dynamic range in the measurement. Nevertheless, low values of B_{IF} increased the acquisition time. As a reference, in [7], the authors used 500 Hz in indoor office channel measurements at mmWave frequencies. Table 1 summarizes the measurement system parameters.

Table 1. Measurement system parameters.

Parameter	Value
VNA output power	−17 dBm
VNA center frequency	26 GHz
VNA SPAN(Bandwidth)	2 GHz
VNA IF Bandwidth (B_{IF})	100 Hz
Frequency points per trace	1091
Tx/Rx antenna gain	5.2 dB
Tx antenna height	0.90 m
Rx antenna height	1.62 m

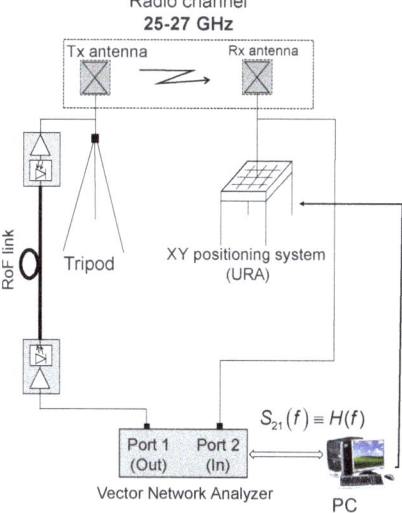

Figure 3. Schematic diagram of the channel sounder.

During the measurements, the Tx antenna was located manually in different locations of the office, imitating the position of user equipment (UE). The Rx subsystem remained fixed in the same position, close to the wall, imitating an access point (AP) that served the users inside the office. The Rx antenna height was 1.62 m with respect to the floor. With the VNA configuration parameters, i.e., N_f and B_{IF}, the acquisition time to capture the CTF in the 144 (12 × 12) positions of the Rx antenna in the URA was about 2 h. To guarantee stationary channel conditions (due to the frequency sweep time to measure the $S_{21}(f)$ scattering parameter, the acquisition of the channel measurements required stationary channel conditions), the measurements were collected at night, thus avoiding the presence of people, not only in the measurement room, but also in adjoining areas. Figure 4 shows the top view of the propagation environment, together with the Rx-URA position and the Tx antenna locations. The channel measurements were performed in LOS and OLOS conditions, defining two scenarios:

- Scenario LOS: The Tx antenna was located at a height of 0.90 m above the floor level, imitating the position of a UE (e.g., a laptop, tablet, or mobile phone) that was on the desk. A total of 10 Tx locations (Tx1–Tx10) was considered in the measurements. Figure 5 (left) shows a view of the Rx-URA and the Tx antenna for the Tx1 position.
- Scenario OLOS: The Tx antenna was also located at a height of 0.90 m above the floor and close to the desk, but in OLOS propagation conditions due to the blockage of the direct component by the computer monitors on the desks. The measurements were taken in 4 Tx locations (Tx11-Tx14). Figure 5 (right) shows a view of the Rx-URA and the Tx antenna for the Tx14 position.

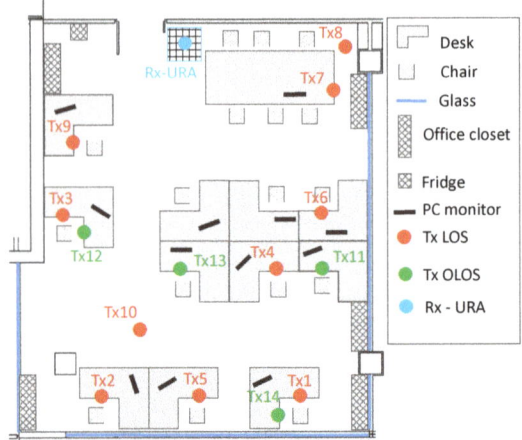

Figure 4. Top view of the propagation environment. The Rx-URA and the Tx antenna locations are indicated.

Figure 5. View of the Rx antenna and the Tx antenna in positions Tx1 in LOS (left) and Tx14 in OLOS (right).

3. Measurement Results

3.1. Path Loss Results

For each position of the URA, the path loss between the Tx and Rx antennas can be derived by averaging the CTF in frequency and taking into account the gain of both antennas in the direct path [13]. Thus, the path loss in logarithmic units, PL, can be derived as (see Appendix A):

$$PL(d) = -10 \log_{10} \left(\frac{1}{N_f} \sum_{n=1}^{N_f} \frac{|H(f_n, d)|^2}{g_{Tx}(f_n) g_{Rx}(f_n) M(f_n)} \right), \tag{1}$$

where d refers to the separation distance between the Tx antenna and the center of the URA for each Tx location, indicated as Tx-Rx distance; f_n is the nth frequency sample; and $g_{Tx}(f_n)$ and $g_{Rx}(f_n)$ are the gain of the Tx and Rx antennas, respectively, in the direction defined by the direct path contribution. The term $M(f_n)$ takes into account the mismatch of the antennas, and it is calculated by:

$$M(f_n) = (1 - |S_{11}^{Tx}(f_n)|^2)(1 - |S_{11}^{Rx}(f_n)|^2), \tag{2}$$

$S_{11}^{Tx}(f_n)$ and $S_{11}^{Rx}(f_n)$ being the $S_{11}(f)$ scattering parameter of the Tx and Rx antennas, respectively.

The measured path loss (cross marker) for each Rx antenna position in the URA in terms of the Tx-Rx distance is shown in Figure 6. Both LOS and OLOS propagation conditions were considered. It is worth noting that the spread values of the path loss along the URA due to the short-term fading was less than 2.5 dB, being less in LOS than in OLOS conditions. It can be observed that the spread values of the path loss did not exhibit any correlation with the Tx-Rx distance. The mean value of the path loss (square marker) for each Tx location is also depicted in Figure 6.

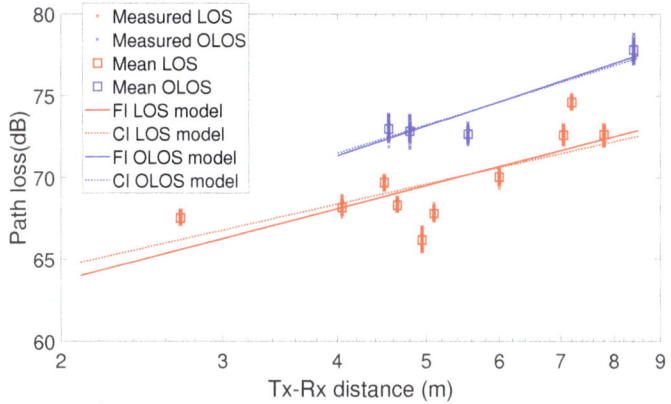

Figure 6. Path loss in terms of the Tx-Rx distance. Measured data, measured mean values, and estimated values from the CI and FI models, in both LOS and OLOS conditions.

The floating-intercept (FI) path loss model has been widely used to describe the behavior of the path loss in terms of the Tx-Rx distance in the microwave frequency band, particularly at the sub-6 GHz band and more recently in mmWave frequencies [5,14], being one of the propagation models adopted in channel standardizations, e.g., the WINNERII Project and 3GPP channel models [15,16]. From the FI model, the path loss is given by:

$$PL^{FI}(d) = \beta + 10\alpha \log_{10}(d) + \chi_\sigma^{FI}, \quad (3)$$

β being the floating-intercept parameter (an offset term); α the path loss exponent, related to both the environment and propagation conditions; and χ_σ^{FI} a zero mean Gaussian random variable, in logarithmic units, with standard deviation σ, which describes the large-scale signal fluctuations about the mean path loss over distance, also known in the literature as the shadow factor (SF). The FI model has a mathematical curve fitting approach over the measured path loss set without any physical anchor.

On the other hand, the close-in (CI) free space reference distance path loss model is also adopted in many studies related to mmWave propagation [5,8]. In the CI model, the path loss is given by:

$$PL^{CI}(d) = FSPL(f, 1\text{ m}) + 10n \log_{10}(d) + \chi_\sigma^{CI}, \quad (4)$$

where $FSPL(f, 1\text{ m}) = 10 \log_{10}(4\pi f/c_0)^2$ is the free space path loss for a Tx-Rx distance equal to 1 m, with c_0 the speed of light; n is the path loss exponent; and χ_σ^{CI} is the SF term. Note that the CI model has certain physical support in the sense that there is an intrinsic frequency dependence of the path loss included in the 1 m FSPL term. Taking into account that $FSPL(f, 1\text{ m})$ is equal to 60.74 dB at 26 GHz, (4) can be rewritten as:

$$PL^{CI}(d) = 60.74 + 10n \log_{10}(d) + \chi_\sigma^{CI}. \quad (5)$$

The path loss fitting results for the FI and CI models are also depicted in Figure 6. Both models exhibit a good fit and predict similar path loss values for the Tx-Rx distance considered, particularly in OLOS conditions. It is worth noting that the maximum path loss difference between LOS and OLOS conditions was about 5 dB, increasing with the Tx-Rx distance. Tables 2 and 3 summarize the mean value and their 95% confidence interval of the FI and CI model parameters. These parameters were derived from the measured path loss using the minimum-mean-squared-error (MMSE) approach.

Table 2. FI path loss model parameters.

	β ($\beta_{95\%}$)	α ($\alpha_{95\%}$)	σ (dB)
LOS	59.29 (58.80–59.79)	1.46 (1.39–1.53)	1.73
OLOS	60.01 (59.46–60.16)	1.88 (1.80–1.95)	0.92

Table 3. CI path loss model parameters.

	FSPL (1 m)	n ($n_{95\%}$)	σ (dB)
LOS	60.74 dB	1.27 (1.26–1.28)	1.75
OLOS	60.74 dB	1.79 (1.78–1.80)	0.93

For the FI model, α had a mean value equal to 1.46, with 1.39–1.53 the 95% confidence interval, in LOS conditions. In OLOS conditions, α had a mean value equal to 1.88, with 1.80–1.95 the 95% confidence interval. For the CI model, α had a mean value equal to 1.27 and 1.79 for LOS and OLOS conditions, respectively. The 95% confidence intervals were narrower for the CI model. In both models, the SF had a similar value, being lower in OLOS conditions.

The values of the path loss exponent derived in this study were lower than the values reported in [8], where path loss exponents in the order or 2.0 and 2.2 were measured at 28 GHz in LOS and non-LOS (NLOS) conditions, respectively, for the FI model. Nevertheless, higher values have been reported for the CI model, where the path loss exponents equal to 1.45 and 2.18 have been measured in LOS and NLOS conditions, respectively. These differences can be explained because the frequencies are slightly different and, of course, due to both the particular characteristics of the environments and propagation conditions. It is worth noting that in our OLOS measurements, only a few MPCs were blocked by the PC monitors, whereas in NLOS conditions, the Tx and Rx were usually separated by different obstructions, and in many cases, the Tx and Rx were not located in the same room. Despite this, our results were more in line with those published by Rappaport et al. in [5,17] for indoor environments at 28 GHz in LOS conditions, where exponents equal to 1.2 and 1.1 were derived for the FI and CI models, respectively, considering omnidirectional path loss modeling (Rappaport et al. used a sliding correlation channel sounder, synthesizing an omnidirectional path loss model from directional measurements). Furthermore, the SF derived was 1.8 dB in both the FI and CI model, a value very close to that obtained by us for the CI model (1.75 dB).

3.2. Time-Dispersion Results

In wideband systems, the multipath propagation causes the arriving signal at the Rx to have a longer duration than the transmitted signal. This effect is the well-known time-dispersion of a wireless channel [11]. In the frequency domain, the time-dispersion can be interpreted as frequency selectivity, i.e., the CTF varies over the bandwidth of interest. The knowledge of the time-dispersion of a wireless channel is vital in the design of wireless systems, for example adopting efficient equalizer structures or defining the optimal multicarrier separation in digital modulations and diversity schemes. In this section, the time-dispersion of the propagation environment is analyzed in both the delay (time) and frequency domains.

3.2.1. Root-Mean-Squared Delay-Spread

The root-mean-squared (rms) delay-spread, denoted by τ_{rms}, is the most relevant parameter to describe the time-dispersion of a wireless channel in the delay domain. τ_{rms} corresponds to the second-order central moment of the power delay profile (PDP), and it is derived as [18]:

$$\tau_{rms}(d) = \sqrt{\frac{\sum_{n=1}^{N_f}(\tau_n - \bar{\tau}(d))^2 PDP(\tau_n, d)}{\sum_{n=1}^{N_f} PDP(\tau_n, d)}}, \quad (6)$$

where τ_n refers to the nth delay bin in the PDP. Assuming ergodicity [11], the PDP can be estimated averaging the channel impulse response (CIR), denoted by $h(\tau, d)$, over all positions of the Rx antenna in the URA:

$$PDP(\tau, d) = E_m\{|h(\tau, d)|^2\}. \quad (7)$$

The CIR is obtained as the inverse Fourier transform of $H(f, d)$. As an example, Figure 7 shows the PDP measured in the Tx1 (LOS condition) and Tx14 (OLOS condition) positions. The differences between LOS and OLOS conditions for low delays can be observed, where the PC monitors blocked the first MPCs, in this case with excess delays around 50 ns.

$$\bar{\tau}(d) = \frac{\sum_{n=1}^{N_f} \tau_n PDP(\tau_n, d)}{\sum_{n=1}^{N_f} PDP(\tau_n, d)}. \quad (8)$$

The cumulative distribution function (CDF) of τ_{rms} for the LOS and OLOS scenarios is shown in Figure 8. A threshold of 30 dB and a Hamming windowing method were considered in the derivation of τ_{rms}. Both curves exhibited a similar trend, with a separation of the order of 3 ns around the mean values. The results showed that the time-dispersion was slightly higher in the OLOS scenario. The minimum, mean, maximum, and standard deviation (STD) values of τ_{rms} are summarized in Table 4. For the LOS scenario, τ_{rms} ranged from 11.21 to 21.74 ns, with a mean value equal to 15.88 ns; whereas in the OLOS scenario, the values were higher, with τ_{rms} ranging from 15.13 to 27.87 ns, with a mean value equal to 18.87 ns, 3 ns more than in the LOS scenario. Nevertheless, the STD of τ_{rms} was very similar in both scenarios, in the order of 2 ns. It is worth noting that the values of τ_{rms} derived here were higher than those published in [7] for an office environment, where values of 8 and 10 ns were reported at 28 and 38 GHz, respectively, in LOS conditions.

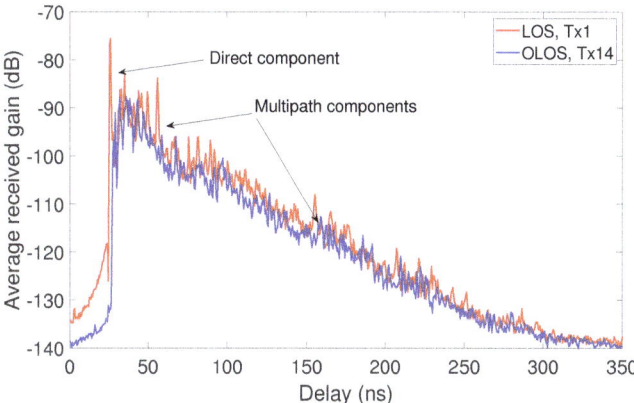

Figure 7. PDP measured in Tx1 and Tx14 positions with LOS and OLOS propagation conditions, respectively.

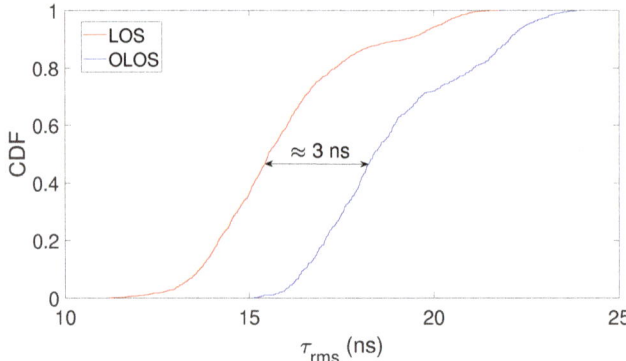

Figure 8. CDF of τ_{rms} for the LOS and OLOS scenarios.

Table 4. Minimum, mean, maximum, and STD values of τ_{rms} (in ns) and $B_{C,90\%}$ (in MHz).

		Minimum	Mean	Maximum	STD
LOS	τ_{rms}	11.21	15.88	21.74	2.01
	$B_{C,90\%}$	3.30	4.88	8.19	0.69
OLOS	τ_{rms}	15.13	18.87	23.87	2.04
	$B_{C,90\%}$	3.02	4.11	5.34	0.51

3.2.2. Coherence Bandwidth

The frequency selectivity of a wideband wireless channel can be described through the frequency correlation function, denoted by $R(\Omega)$, calculated as the Fourier transform of the PDP [18]:

$$R(\Omega) = \int_0^\infty PDP(\tau) \exp(-j2\pi\Omega\tau) d\tau. \tag{9}$$

From (9), the coherence bandwidth can be defined as the smallest value of Ω for which $R(\Omega)$ equals a certain correlation coefficient, typically 0.9 (or 90%). Figure 9 shows the CDF of the 90% coherence bandwidth, denoted as $B_{C,90\%}$. The minimum, mean, maximum, and standard deviation (STD) values of $B_{C,90\%}$ are summarized in Table 4. The values of the coherence bandwidth had a higher dispersion in LOS conditions, where $B_{C,90\%}$ ranged from 3.30 to 8.19 MHz, with a mean value equal to 4.88 MHz. In OLOS conditions, the maximum value of $B_{C,90\%}$ was in the order of 5 MHz. Nevertheless, the difference between the mean value in LOS and OLOS conditions was not significantly high, about 0.8 MHz.

In order to investigate the relationship between the time-dispersion in the delay and frequency domains, Figures 10 and 11 show the scatter plot of the 90% coherence bandwidth versus τ_{rms}. The black line in both figures establishes a relationship between $B_{C,90\%}$ and τ_{rms} in the form:

$$B_{C,90\%}(\text{MHz}) = \frac{\alpha_0}{\tau_{rms}(\text{ns})^\gamma}. \tag{10}$$

Equation (10) is an empirical expression, first proposed in [19], to try to model the relationship between the coherence bandwidth and τ_{rms}. The relationship between the PDP and the autocorrelation function through the Fourier transform makes the consistent assumption that the relationship between $B_{C,90\%}$ and τ_{rms} must be inverse. In fact, prior to the model given by (10), simpler expressions have been proposed, defined by a single parameter, α_0, considering $\gamma = 1$ in (10) [20]. The variability of the PDP in different locations, even in the same propagation environment, makes it advisable to adopt a two-parameter model as proposed in [19]. Table 5 summarizes the values of α_0 and γ that appear in (10) using the MMSE approach. The 95% confidence intervals are also included in the table. Notice that the mean value of γ was the same in both scenarios; only differences appear in the values that define the 95% confidence interval.

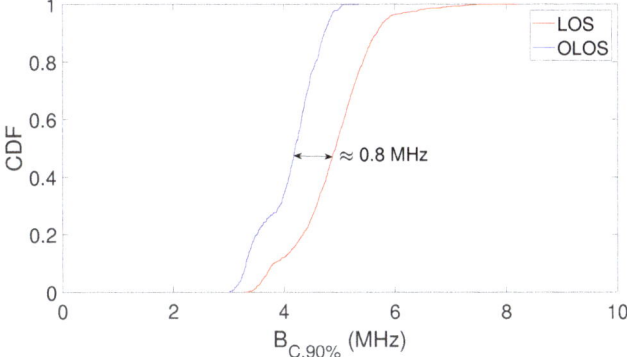

Figure 9. CDF of $B_{C,90\%}$ for the LOS and OLOS scenarios.

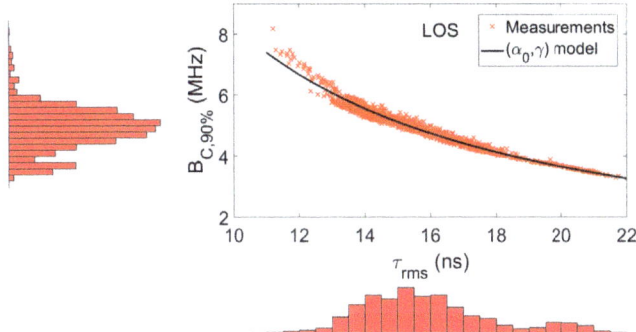

Figure 10. Scatter plot of the 90% coherence bandwidth versus τ_{rms} in LOS conditions.

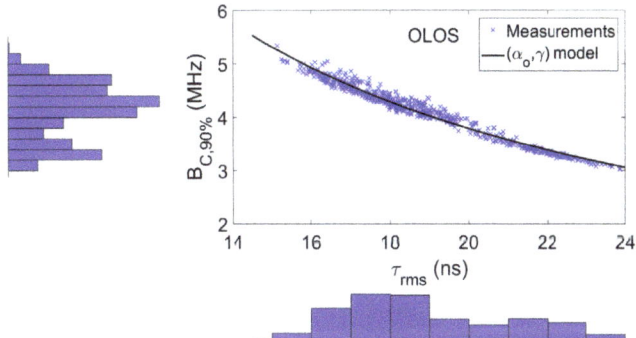

Figure 11. Scatter plot of the 90% coherence bandwidth versus τ_{rms} in OLOS conditions.

Table 5. Values of α_0 and γ (in ns).

	$\alpha_0\,(\alpha_{0,95\%})$	$\gamma\,(\gamma_{95\%})$
LOS	124.5 (120.8–128.2)	1.178 (1.167–1.189)
OLOS	129.2 (122.2–136.2)	1.178 (1.160–1.197)

4. Conclusions

In this paper, the path loss and time-dispersion characteristics of the propagation channel in a typical office environment were analyzed from a channel measurement campaign carried out at 26 GHz. The channel measurements were collected under LOS and OLOS conditions. The parameters of the FI and CI path loss models and their 95% confidence interval were derived from the measurements using the MMSE approach. Mean path loss exponents equal to 1.46 and 1.88 were derived for the FI model in LOS and OLOS, respectively. For the CI model, the mean path loss exponent was equal to 1.27 and 1.79 in LOS and OLOS, respectively. The results showed a maximum path loss difference between LOS and OLOS in the order of 5 dB.

The time-dispersion due to the multipath propagation was analyzed through the delay-spread and the coherence bandwidth. Mean values of τ_{rms} equal to 15.88 and 18.87 ns were derived in LOS and OLOS, respectively; and mean values of the 90% coherence bandwidth equal to 4.88 and 4.11 were derived in LOS and OLOS, respectively. Furthermore, the correlation between the coherence bandwidth and the delay-spread was investigated, observing an inverse relationship between them.

The results reported in this contribution enable better knowledge of the propagation characteristics in office environments at 26 GHz and can be used to improve the design and evaluate the performance of the future 5G networks in these scenarios.

Author Contributions: Conceptualization, L.R., R.P.T., V.M.R.P. and J.-M.M.-G.-P.; data curation, L.R., V.M.R.P., J.R. and J.-M.M.-G.-P.; formal analysis, L.R., R.P.T., V.M.R.P. and J.-M.M.-G.-P.; funding acquisition, L.R., R.P.T., V.M.R.P. and J.-M.M.-G.-P.; investigation, L.R., R.P.T., V.M.R.P., J.R.P., J.R. and J.-M.M.-G.-P.; methodology, L.R., V.M.R.P. and R.P.T.; visualization, L.R., V.M.R.P., J.R. and H.F.; writing, review and editing, L.R., R.P.T., V.M.R.P., J.R.P., H.F., J.-M.M.-G.-P. and J.R.

Funding: This work was funded in part by the Spanish Ministerio de Economía, Industria y Competitividad under the research projects TEC2016-78028-C3-2-P, TEC2017-86779C1-2-R, and TEC2017-86779-C2-2-R, and by COLCIENCIAS in Colombia.

Acknowledgments: The authors want to thank Víctor Rubio for his support during the measurement campaign, as well as Bernardo Bernardo-Clemente for his support and assistance in the laboratory activities.

Conflicts of Interest: The authors declare no conflict of interest.

Abbreviations

The following abbreviations are used in this manuscript:

3D	Three-dimensional
5G	Fifth-generation
AP	Access point
CDF	Cumulative distribution function
CI	Close-in free space reference distance path loss model
CIR	Channel impulse response
CTF	Channel transfer function
DUT	Device under test
FI	Floating-intercept path loss model
HPBW	Half power beamwidth
ITU	International Telecommunication Union
LOS	Line-of-sight

mmWave	Millimeter wave
MPC	Multipath contribution
OLOS	Obstructed-LOS
PDP	Power delay profile
RoF	Radio over fiber
RSPG	Radio Spectrum Policy Group
Rx	Receiver
STD	Standard deviation
Tx	Transmitter
UE	User equipment
URA	Uniform rectangular array
VNA	Vector network analyzer
WRC	World Radiocommunication Conference

Appendix A. Path Loss Derivation

From the Friis transmission formula for a narrowband system, the relationship between the received power, denoted by $P_r(f)$, and the transmit power, denoted by to $P_t(f)$, is expressed as:

$$\frac{P_r(f)}{P_t(f)} = \frac{g_{Tx}(f)g_{Rx}(f)}{L_{FS}(f,d)}, \tag{A1}$$

d being the Tx-Rx distance; $L_{FS}(f,d) = (4\pi df/c_0)^2$ the free space path loss, with c_0 the speed of light; $g_{Tx}(f)$ and $g_{Rx}(f)$ the gain of the Tx and Rx antennas in the direct path, respectively. In multipath propagation, (A1) can be completed by a term $A_{MPC}(f,d)$:

$$\frac{P_r(f)}{P_t(f)} = \frac{g_{Tx}(f)g_{Rx}(f)}{L_{FS}(f,d)} A_{MPC}(f,d). \tag{A2}$$

Notice that the term $A_{MPC}(f,d)$ in (A2) takes into account the interference effect of the MPCs at the receiver.

Let us define the propagation gain, denoted by $G(f,d)$, as:

$$G(f,d) = L_{FS}^{-1}(f,d) A_{MPC}(f,d). \tag{A3}$$

Then, (A2) can be written as:

$$\frac{P_r(f)}{P_t(f)} = g_{Tx}(f)g_{Rx}(f)G(f,d). \tag{A4}$$

Now, taking into account that there may be a mismatch in the Tx and Rx antennas, this effect can be modeled by a term $M(f)$ calculated by:

$$M(f) = (1 - |S_{11}^{Tx}(f)|^2)(1 - |S_{11}^{Rx}(f)|^2), \tag{A5}$$

$S_{11}^{Tx}(f)$ and $S_{11}^{Rx}(f)$ being the $S_{11}(f)$ scattering parameter of the Tx and Rx antennas, respectively. Thus, the relationship between $P_r(f)$ and $P_t(f)$ can be written in a general form as:

$$\frac{P_r(f)}{P_t(f)} = g_{Tx}(f)g_{Rx}(f)G(f,d)M(f). \tag{A6}$$

From (A6), the propagation gain can be expressed as:

$$G(f,d) = \frac{P_r(f)}{P_t(f)} \frac{1}{g_{Tx}(f)g_{Rx}(f)M(f)}. \tag{A7}$$

Taking into account that the received and transmit power are related to the $S_{21}(f,d)$ scattering parameter measured by the VNA,

$$|S_{21}(f,d)|^2 = \frac{P_r(f)}{P_t(f)}, \tag{A8}$$

and that $S_{21}(f,d)$ is equivalent to the CTF, i.e., $H(f,d) \equiv S_{21}(f,d)$, from (A7), the propagation gain can be expressed as:

$$G(f,d) = \frac{|H(f,d)|^2}{g_{Tx}(f)g_{Rx}(f)M(f)}. \tag{A9}$$

For a wideband system, the transmission gain or path gain, denoted by $PG(d)$, is defined as:

$$PG(d) = E_f\{G(f,d)\}, \tag{A10}$$

where $E_f\{\cdot\}$ denotes expectation over the frequency bandwidth. Thus, from channel measurements, the path gain can be estimated as follows:

$$PG(d) = \frac{1}{N_f}\sum_{n=1}^{N_f}\frac{|H(f_n,d)|^2}{g_{Tx}(f_n)g_{Rx}(f_n)M(f_n)}, \tag{A11}$$

where f_n is the nth frequency sample and N_f is the number of frequency points over the measured bandwidth. Finally, the path loss is defined as the inverse of the path gain. Then, from (A11), the path loss in logarithmic units, denoted by $PL(d)$, can be derived as:

$$PL(d) = -10\log_{10}\left(\frac{1}{N_f}\sum_{n=1}^{N_f}\frac{|H(f_n,d)|^2}{g_{Tx}(f_n)g_{Rx}(f_n)M(f_n)}\right), \tag{A12}$$

References

1. Samsung R&D. 5G Vision. 2015. Available online: https://images.samsung.com/is/content/samsung/p5/global/business/networks/insights/white-paper/5g-vision/global-networks-insight-samsung-5g-vision-2.pdf (accessed on 24 September 2019).
2. Andrews, J.G.; Buzzi, S.; Choi, W.; Hanly, S.V.; Lozano, A. What will 5G be? *IEEE J. Sel. Areas Commun.* **2014**, *32*, 1065–1082. [CrossRef]
3. European Commission—Radio Spectrum Policy Group. Strategic Roadmap Towards 5G for Europe. 2016. Available online: https://rspg-spectrum.eu/wp-content/uploads/2013/05/RPSG16-032-Opinion_5G.pdf (accessed on 24 September 2019).
4. Resolution 238. In Proceedings of the World Radio Communications Conference, Geneva, Switzerland, 2–27 November 2015.
5. MacCartney, G.R.; Rappaport, T.S.; Sun, S.; Deng, S. Indoor office wideband millimeter-wave propagation measurements and channel models at 28 GHz and 73 GHz for ultra-dense 5G wireless networks (invited paper). *IEEE Access* **2015**, *3*, 2388–2424. [CrossRef]
6. Haneda, K.; Järveläinen, J.; Karttunen, A.; Kyrö, M.; Putkonen, J. A statistical spatio-temporal radio channel model for large indoor environments at 60 and 70 GHz. *IEEE Trans. Antennas Propag.* **2015**, *63*, 2694–2704. [CrossRef]
7. Huang, J.; Wang, C.X.; Feng, R.; Zhang, W.; Yang, Y. Multifrequency mmWave massive MIMO channel measurements and characterization for 5G wireless communications systems. *IEEE J. Sel. Areas Commun.* **2017**, *35*, 1591–1605. [CrossRef]
8. Tang, P.; Zhang, J.; Shafi, M.; Dmochwski, P.A.; Smith, P.J. Millimeter wave channel measurements and modeling in an indoor hotspot scenario at 28 GHz. In Proceedings of the 2018 IEEE 88th Vehicular Technology Conference (VTC-Fall), Chicago, IL, USA, 27–30 August 2018; pp. 1–5.

9. Rodrigo-Peñarrocha, V.M.; Rubio, L.; Reig, J.; Juan-Llácer, L.; Pascual-García, J.; Molina-Garcia-Pardo, J.M. Millimeter wave channel measurements in an intra-wagon environment. In Proceedings of the COST CA15104 TD(18)07040, Cartagena, Spain, 29 May 2018; pp. 1–5.
10. Sánchez, M.G.; de Haro, L.; Pino, A.G.; Calvo, M. Human operator effect on wide-band radio channel characteristics. *IEEE Trans. Antennas Propag.* **1997**, *45*, 1318–1320. [CrossRef]
11. Molisch, A.F. *Wireless Communications*, 2nd ed.; Wiley-IEEE Press: Chichester, UK, 2010.
12. Steele, R.; Hanzo, L. *Mobile Radio Communications*, 2nd ed.; Wiley: Chichester, UK, 1999.
13. Rubio, L.; Reig, J.; Fernández, H.; Rodrigo-Peñarrocha, V.M. Experimental UWB propagation channel path loss and time-dispersion characterization in a laboratory environment. *Int. J. Antennas Propag.* **2013**, *2013*, 35017. [CrossRef]
14. Deng, S.; Samimi, M.K.; Rappaport, T.S. 28 GHz and 73 GHz millimeter-wave indoor propagation measurements and path loss models. In Proceedings of the IEEE International Conference on Communications, London, UK, 8–12 June 2015; pp. 1244–1250.
15. Meinilä, J.; Kyösti, P.; Jämsä, T.; Hentilä, L. *WINNER II Channel Models*; IST-4-027756-WINNER, Tech. Rep. D1.1.2; Inf. Soc. Technol, 2007. Available online: https://www.cept.org/files/8339/winner2%20-%20final%20report.pdf (accessed on 24 September 2019).
16. 3GPP TR 25.996. *Spatial Channel Model for Multiple Input Multiple Output (MIMO) Simulations*; ETSI Technical Repport 125 996; ETSI: Sophia Antipolis, France, 2012.
17. Sun, S.; MacCartney, G.R.; Rappaport, T.S. Millimeter-wave distance-dependent large-scale propagation measurements and path loss models for outdoor and indoor 5G systems. In Proceedings of the 2016 10th European Conference on Antennas and Propagation (EuCAP), Davos, Switzerland, 10–15 April 2016; pp. 1–5.
18. Parsons, J.D. *The Mobile Radio Propagation Channel*, 2nd ed.; Wiley: Chichester, UK, 2000.
19. Howard, S.J.; Pahlavan, K. Measurements and analysis of the indoor radio channel in the frequency domain. *IEEE Trans. Instrum. Meas.* **1990**, *39*, 751–755. [CrossRef]
20. Gans, M. A power-spectral theory of propagation in the mobile radio environment. *IEEE Trans. Veh. Technol.* **1972**, *21*, 27–37. [CrossRef]

© 2019 by the authors. Licensee MDPI, Basel, Switzerland. This article is an open access article distributed under the terms and conditions of the Creative Commons Attribution (CC BY) license (http://creativecommons.org/licenses/by/4.0/).

Article

Wideband Performance Comparison between the 40 GHz and 60 GHz Frequency Bands for Indoor Radio Channels

Miguel Riobó [1], Rob Hofman [1,2], Iñigo Cuiñas [1], Manuel García Sánchez [1,*] and Jo Verhaevert [2]

[1] Signal Theory and Communications Department, University of Vigo, 36310 Vigo, Spain; miriobo@uvigo.es (M.R.); rob.hofman@ugent.be (R.H.); inhigo@uvigo.es (I.C.)
[2] Department of Information Technology, Ghent University/imec, 9000 Gent, Belgium; Jo.Verhaevert@UGent.be
* Correspondence: manuel.garciasanchez@uvigo.es; Tel.: +34-986812195

Received: 27 September 2019; Accepted: 26 October 2019; Published: 29 October 2019

Abstract: When 5G networks are to be deployed, the usability of millimeter-wave frequency allocations seems to be left out of the debate. However, there is an open question regarding the advantages and disadvantages of the main candidates for this allocation: The use of the licensed spectrum near 40 GHz or the unlicensed band at 60 GHz. Both bands may be adequate for high performance radio communication systems, and this paper provides insight into such alternatives. A large measurement campaign supplied enough data to analyze and to evaluate the network performance for both frequency bands in different types of indoor environments: Both large rooms and narrow corridors, and both line of sight and obstructed line of sight conditions. As a result of such a campaign and after a deep analysis in terms of wideband parameters, the radio channel usability is analyzed with numerical data regarding its performance.

Keywords: 5G; radio channel; measurements; wideband; indoor

1. Introduction

A society without wireless communication is unthinkable nowadays. In the latest decades, people and enterprises have used wireless communication every day and the evolution in wireless technologies has redefined how people work and communicate. Beside consumers, enterprises have also embraced this revolution to work faster and more efficiently.

In 1982, the first generation of mobile communications (1G) was released: Analog-based and very limited, as voice service was its only capability. Ten years later, the digital version appeared (2G), offering the possibility of sending data through the network in the form of what was called short message system (SMS). The following generation, 3G, along with the then recently introduced smartphones, offered the possibility of sharing multimedia data, faster web browsing and even video calling. The fourth generation, 4G, increased the performance of the previous generation in terms of data rate. The next revolution in wireless communication is right around the corner: The fifth generation (5G). This technology promises higher speeds and lower latencies. To accommodate this new generation, frequency bands both below and over current 4G ones will be used [1]. Nowadays, with the emergence of streaming platforms allowing via a mobile connection to watch movies and series or even to play games, the consumer is demanding even more data rate in order to be able to keep up with the increasing quality of the content. 5G will try to provide a solution to all these demands and even more, like the low latency required in vehicle to vehicle or vehicle to infrastructure communications for autonomous cars.

When designing 5G, some doubts arose on the usage of millimeter wave frequency bands. Rappaport et al. supported that 5G would get consistent coverage at 38 GHz by installing base stations with a 200-meter cell-radius in dense urban environments [2]. Once any worry related to the usability of such frequencies was removed, a normalization process was started by consortia (like 3GPP) and institutions (like ITU). Today some public administrations (like FCC) have announced auctions of spectrum parts at millimeter wave frequencies for 5G [3]. Alternatively, there is an unlicensed band around 60 GHz that is also well positioned to succeed as a candidate for providing high performance services based on radio communication systems [4]. Both frequency bands allow high rate traffic and the final decision depends on a deep analysis of their advantages and disadvantages.

Facing the 5G expectation, a precise channel characterization is necessary, as network planners are concerned on achieving the required quality of service. However, indoor scenarios still need deeper analysis to assure that propagation within such environments fits within the expected performance of those frequency bands. Thus, this paper proposes an experimental analysis from the results of a measurement campaign at two frequency channels in the millimeter wave range: 40 GHz and 60 GHz. The measurements are performed in a corridor in both a line of sight (LOS) and obstructed line of sight (OLOS) scenarios, and in an ordinary auditorium used for teaching activities.

As a result, this paper provides wideband propagation parameters at both frequency bands and in different environments, giving an insight into their characteristics and also helping network planners in selecting the most adequate frequency channels. We detected that the effect of multipath would be less important in the 60 GHz band, as attenuations by propagation mechanisms will reduce the influence of contributions with larger delays. In general, the root mean square (RMS) delay spread is larger at 40 GHz than at 60 GHz, which could limit the usability for wideband systems. The worst case also occurs in OLOS conditions, which cannot be neglected in a real world indoor radio link.

After this introduction, the setup of the measurements is discussed in Section 2, devoted to materials and methods. This section also contains the description of the environments, the post processing techniques and the ray-tracing analysis used to check the origin of the observed effects. Section 3 contains the wideband results at both the corridor and the auditorium environments. This section relates the power delay profile (PDP) to the shape of the environment, with the help of the ray-tracing analysis, and it also shows the main wideband parameters as time spread and coherence bandwidth. The discussion of these results is handled in Section 4. Finally, Section 5 summarizes the conclusions.

2. Materials and Methods

This section describes the measurement, analysis and interpretation of the collected data: It starts with the measurement setup and procedure, it continues with the measurement environments and the post processing applied to the raw data. It finishes with the ray-tracing analysis used to explain the different observed effects within the indoor environments.

2.1. Measurement Setup and Procedure

Measurements followed the frequency swept method [5]. A Rhode & Schwarz ZVA67 vector network analyzer (VNA) (Rohde & Schwarz, Munich, Germany) executed the measurements of the frequency response of the radio channel. This equipment transmits a frequency-swept signal in the band of interest and gathers the corresponding received signal at the other ends. It also computes the S-parameters of the measured channel as a function of the frequency. To measure at 40 GHz, the transmitter was connected to an amplifier through a 4 m coaxial cable and then to the antenna through a 1 m coaxial cable, making this way a total length of 5 m. At 60 GHz only the 4 m cable was necessary, as the amplifier was a wave guide one and was connected directly to the transmitting antenna. Furthermore, the receiver end consisted of three omnidirectional antennas connected to three VNA receiving ports through three 2 m coaxial cables. Despite future mobile terminals working at millimeter wave frequencies will probably use directive antennas that will filter some multipath components; by using omnidirectional ones in our measurement system we will be able to get a better idea about all the multipath components present in the environments.

The measured frequency bands extend 3 GHz around 41.5 GHz and 60.5 GHz. Each complex frequency response comprises 1001 equally separated frequency samples resulting in a frequency resolution of 3 MHz. A thru calibration of the measurement setup was performed before the field measurements were taken.

The transmitting antenna is attached to the ceiling, whereas the three receiving antennas are mounted close to each other on a stand placed over a straight-line automatized rail, as depicted in Figure 1. Those antennas are mounted in an equilateral triangle with sides of 4 cm for the 40 GHz case and 2.5 cm for the 60 GHz one, following the scheme of Figure 2. These distances are equivalent to 5.3 wavelengths at 40 GHz and 5 wavelengths at 60.5 GHz. The measured mutual coupling shows values below −35 dB and −25 dB at 40 GHz and 60.5 GHz, respectively. The antennas were placed 77 cm above the floor. The antenna labeled as "3" was the closest to the transmitter; the other two are labeled as "2" and "4". Note that the number of the antenna corresponds with the S parameter measured on the VNA, considering label "1" of the transmitter antenna. For example, S_{31} corresponds to the channel response of receiver antenna number "3".

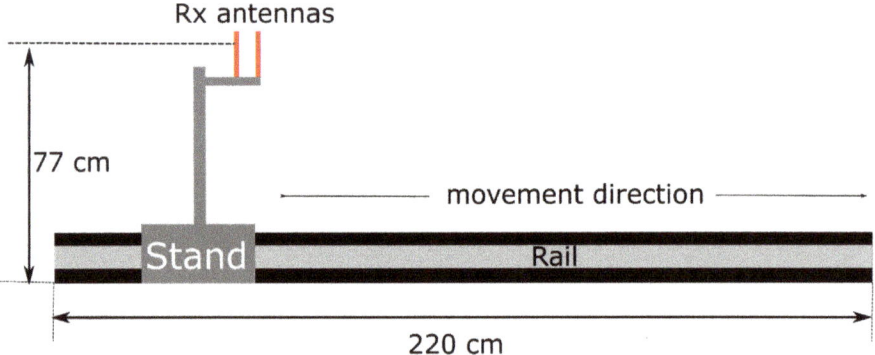

Figure 1. Measurement setup of the receiver.

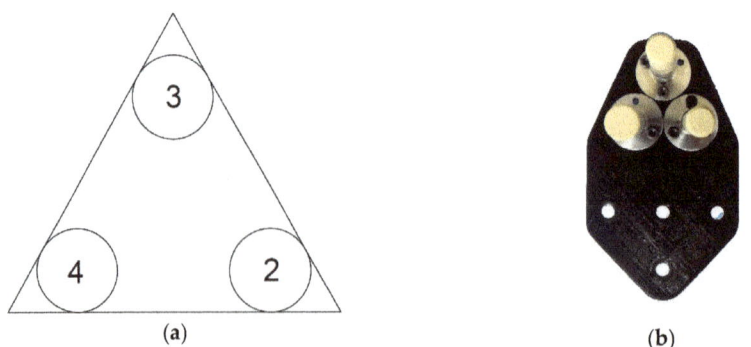

Figure 2. Triangular placement of three receiving antennas at 60 GHz band (**a**) scheme; (**b**) final realization.

The receiver antennas move over a certain distance on the rail, controlled by an indexer and a laptop. The movement of the stand occurs along 220 cm, in steps of 1.875 mm for the 40 GHz channel and 1.25 mm for the 60 GHz channel. These step lengths are a quarter of the wavelength of their corresponding center frequency. Measurements are repeated with the rail placed at different positions along the corridor, enlarging the entire measurements longer than the length of the rail itself.

A tailor-made Matlab program was developed to capture and process the data. This program runs on a laptop connected to the indexer controlling via a serial RS232 connection the step-by-step motor of the rail and to the VNA using TCP/IP over a UTP cable.

2.2. Measurement Environments

Measurements took place in two different scenarios: An ordinary auditorium used for teaching activities and a corridor, both at the School of Telecommunication Engineering at the University of Vigo. During measurements the presence of people moving within the environments was not allowed, only the antennas changed their position. This results in a static environment.

The auditorium is a flat classroom furnished with chairs and tables divided into four big blocks. It has a brick wall covered by plaster on one of the sides, while the opposite wall has plenty of large windows. At the back of the auditorium there is another wall with a couple of glass doors and windows. The last wall has a blackboard and some more windows. The transmitting antenna was located at the ceiling, near the front of the room, and the three receiver antennas were on the moving stand of the rail. Figure 3 depicts this first environment, including the location of transmitting antenna and the paths for the receiving antenna during measurements.

(a)

Figure 3. *Cont.*

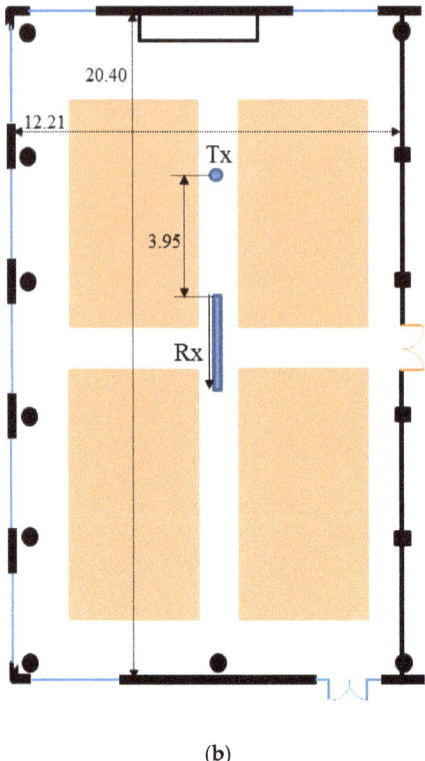

(b)

Figure 3. Description of the auditorium environment: (**a**) Picture; (**b**) floor plan with dimensions in m.

The second scenario is a narrow corridor, with brick walls covered by plaster. The floor plans are depicted in Figure 4. This scenario hosted two different environments: LOS and OLOS, depending on the location of the transmitting antenna and its visibility from the receiving ones. The receiving antenna paths were at the exact center of the 2.20-m width corridor, whereas the transmitting antenna location varied depending on the configured environment: For LOS the antenna was at the ceiling, also in the middle of the corridor; for the OLOS, a wall corner obstructed the line of sight between the antenna (again at the ceiling) and the receiving path. Figure 5 shows a picture taken during the measurements.

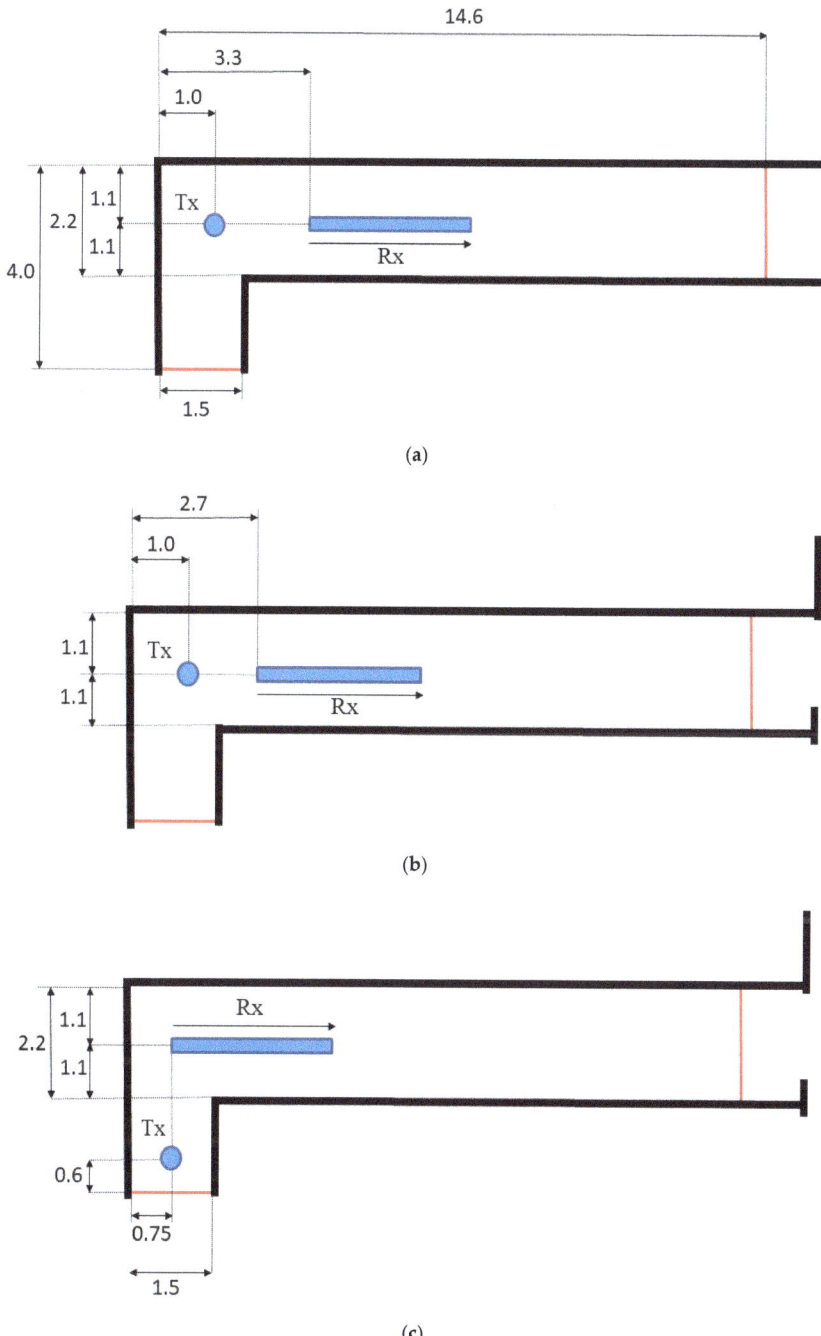

Figure 4. Floor plans of the corridor, all dimensions in meters; (**a**) 40 GHz LOS, (**b**) 60 GHz LOS, (**c**) OLOS.

Figure 5. Picture of the corridor during the measurement campaign.

2.3. Data Processing

An inverse Fourier transform is applied to the measured complex frequency responses of the radio channel in order to obtain the channel impulse responses $h(t,\tau)$. Due to the limited frequency range of the measurements the frequency response is windowed and the impulse response may contain certain peaks originating from the side lobes of the window response in the time domain. To remove these peaks which do not correspond to actual propagation paths, as well as part of the noise, we set a threshold according to [6] and remove any sample below that threshold. Then we use those impulse responses to compute the power delay profile (PDP) as is shown in Equation (1) and represent the power received at a certain location as a function of the delay, τ [7,8].

$$PDP(\tau) = \frac{1}{N} \sum_{n=1}^{N} |h(\tau, n \cdot \Delta l)|^2 \qquad (1)$$

where Δl is the step distance the receiver antennas are moved between successive measurements.

As seen in Equation (1), there is an averaging process of N impulse responses to obtain the PDP. We used a sliding average with N = 11 to obtain the results of this paper [7]. Based on the PDP vectors some relevant time dispersion parameters were computed: The mean delay and the RMS delay spread. The mean delay is the first moment of the power delay profile, and it is defined as in Equation (2).

$$\tau_{mean} = \frac{\sum_{k=0}^{\infty} PDP(\tau_k) \tau_k}{\sum_{k=0}^{\infty} PDP(\tau_k)} \qquad (2)$$

The RMS delay spread is even more interesting than the mean delay, as it is related to the inter-symbol interference (ISI) phenomenon giving cause to the irreducible bit error rate (BER) of the radio communication system. The RMS delay spread is the standard deviation of the delay of multipath components, weighted proportional to the energy transported by each of them. This RMS delay spread is also the square root of the second central moment of the PDP. The formula for the RMS delay spread is shown in Equation (3) [8,9]. The symbol interval used in the system has to be much larger than the RMS delay spread of the radio channel to avoid ISI [10].

$$\tau_{rms} = \sqrt{\frac{\sum_{k=0}^{\infty}(\tau_k - \tau_{mean})^2 PDP(\tau_k)}{\sum_{k=0}^{\infty} PDP(\tau_k)}} \quad (3)$$

Both time dispersion and frequency selectivity are important factors when dealing with multipath channels. The most commonly used parameter to express frequency selectivity is the coherence bandwidth (Bc). This bandwidth is a statistical parameter, and can be defined as that part of the bandwidth where the channel response is considered flat. In other words, it represents the range of frequencies that experience correlated fading [11,12].

Computing the frequency correlation function (FCF or $R\tau$) is the first step to calculate the coherence bandwidth, Bc. The frequency correlation function is the Fourier transform of the PDP, as indicated in Equation (4). Typical values of the correlation level (α) used to estimate Bc are 0.5, 0.7 and 0.9. The Bc is then that part of the frequency range where the FCF is above one of these levels [8,11,12].

$$R_\tau(\Delta f) = \int_{-\infty}^{\infty} PDP(\tau) e^{-j2\pi\Delta f\tau} d\tau \quad (4)$$

Figure 6 depicts the estimation of coherence bandwidth at those correlation levels on the computed frequency correlation function for the auditorium at 40 GHz and received by antenna "3". The coherence bandwidths at which the frequency correlation function is above 0.9, 0.7 and 0.5, are, respectively 24 MHz, 138 MHz and 222 MHz.

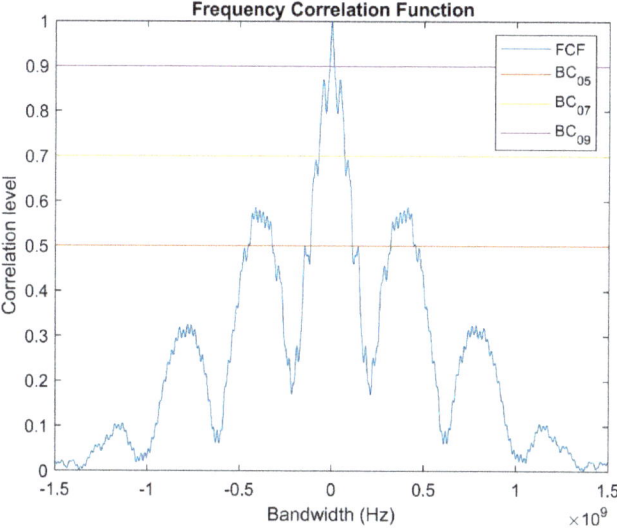

Figure 6. Frequency correlation function and estimation of coherence bandwidth at 40 GHz in the auditorium.

2.4. Ray-Tracing Analysis

The time difference in the various echoes relative to the first arriving component (the one corresponding to the direct path in an LOS situation) can give some information about where they have been produced. Multipath components in the PDP are associated with a certain delay and thus with a certain travelled distance. Using these delays and the dimensions and geometry of the environment, each multipath component can be mapped to a certain scattering element in the environment, wherefore a very simple ray-tracing tool has been developed only calculating the origin of the different multipath components and not the amplitude of them. Using image theory [13], first order reflections on the most significant elements were computed.

Figure 7 shows, as an example, the fit between the PDP obtained from measurements and the delay of different propagation paths computed by the above-described ray-tracing tool. Up to four different paths considered by the ray-tracing tool superimpose to the measured PDP, whereas the LOS component and the floor reflection fit perfectly the empirical results, the reflection caused by the back door seems just a bit ahead of the measured one. This may be due to small measurement errors, a more complex multipath propagation than that used in the simulations, or even some additional delaying elements not considered in this analysis.

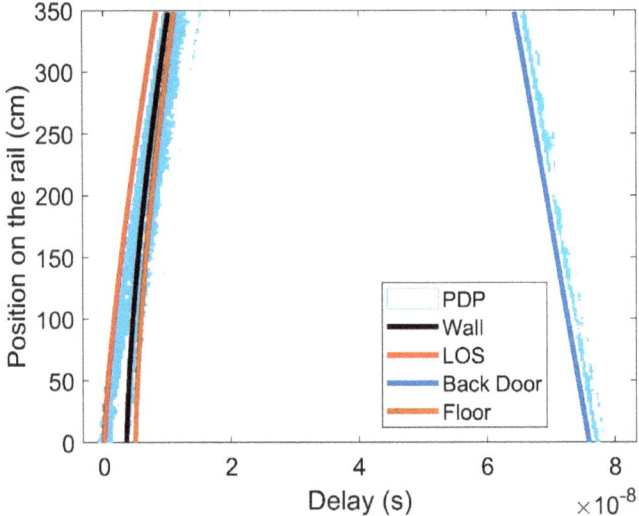

Figure 7. Mapping the measured power delay profile (PDP) and ray-tracing estimations for the corridor channel received by antenna "3" in line of sight (LOS) conditions at the 60 GHz band.

3. Results

In this section, results obtained in the auditorium and the corridor, are presented. Every measurement results in three different S-parameters, measured at the same time, which allows observing the difference between multipath components arriving at the three antennas. All wideband results are shown, relative to the delay of the main LOS component.

3.1. Auditorium

Figure 8a–c show the PDP for the three S-parameters at 41.5 GHz. Taking a closer look at the area of the component arriving first, it is easy to see that antenna "2" (closest to the transmitter) gets the most energy of the three. This can be explained by the fact that also reflections from the floor, tables and chairs are taken together, where those are at the same time partially blocked to the other antennas. Another explanation can be the coupling of the different receiving antennas, resulting in (smaller) modifications of the radiation pattern and hence reducing efficiency and performance.

The direct component can be easily identified as the one with the smallest delay. As the receivers move away from the transmitter, that relative delay increases. The same trend can be seen in other contributions produced by scatterers placed between the transmitter and the receivers.

However, at the central part of the graphs, there is a contribution that can be seen in Figure 8a (and less clearly in Figure 8c) starting at 20 ns delay and lowering its delay as the receivers move away from the transmitter. This indicates that the contribution is arriving from a direction opposite to that of the direct component. It was probably produced by the laptop, which was the closest to antenna "2".

As our simple ray-tracing tool does not take into account the amplitude of the components, it could predict some components that do not appear in the actual measured results due to their large attenuation and low power in the real world. This is the case for the sidewall reflection in Figure 8.

Finally, the contributions coming from the chairs at the back area of the auditorium can also be identified by their initially longer delays, getting shorter when the receiver moves closer to these scatterers.

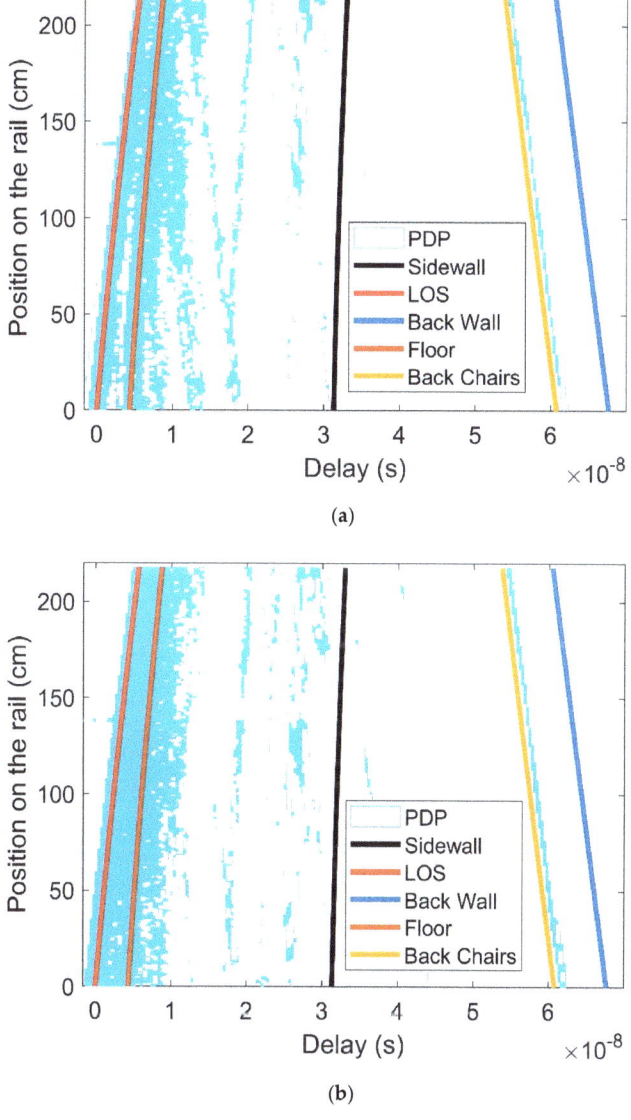

Figure 8. Cont.

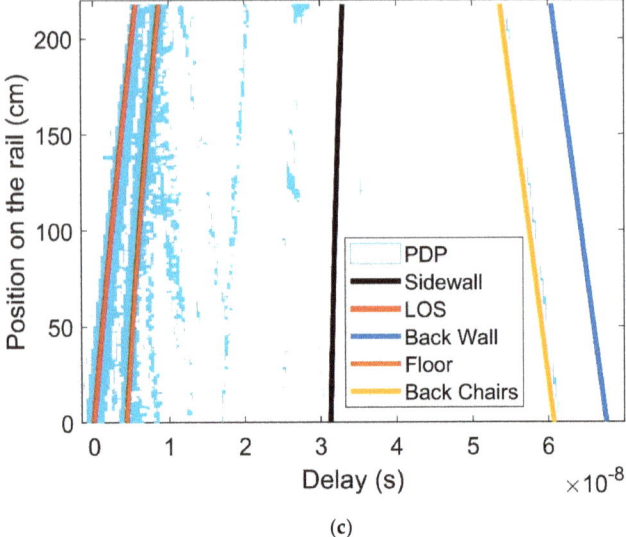

(c)

Figure 8. PDP at auditorium, at 41.5 GHz, moving the receiver along the central corridor of the auditorium: (**a**) Received by antenna "2"; (**b**) by antenna "3"; (**c**) by antenna "4".

Figure 9 shows the measured PDP at the 60 GHz band together with the ray-tracing predictions. Aside from the difference between the three receivers, it is also interesting to recall the differences on the PDP between 40 GHz and 60 GHz. While for 40 GHz there are several multipath components, in 60 GHz only the line of sight and the reflection on the floor are present. The other contributions from the side and back wall suffer stronger attenuations and fall below the receiver's sensitivity (Figure 9a–c). In fact, the ray-tracing simulations predict the time delays that would correspond to those contributions, as can be seen in Figure 10, but no signal over the noise threshold was detected in the measurements.

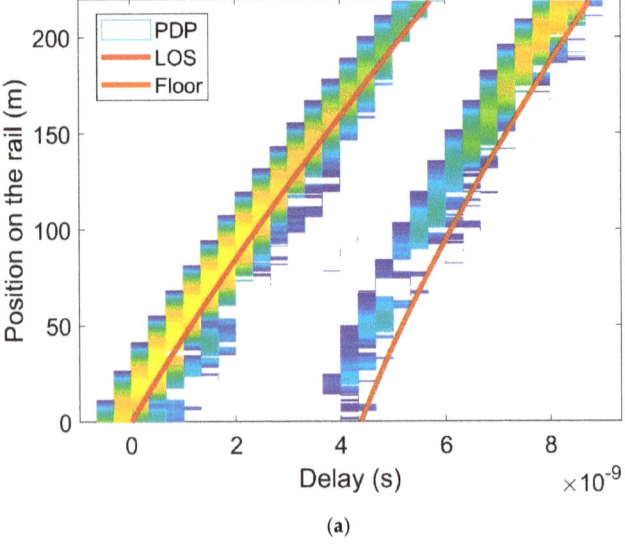

(a)

Figure 9. *Cont.*

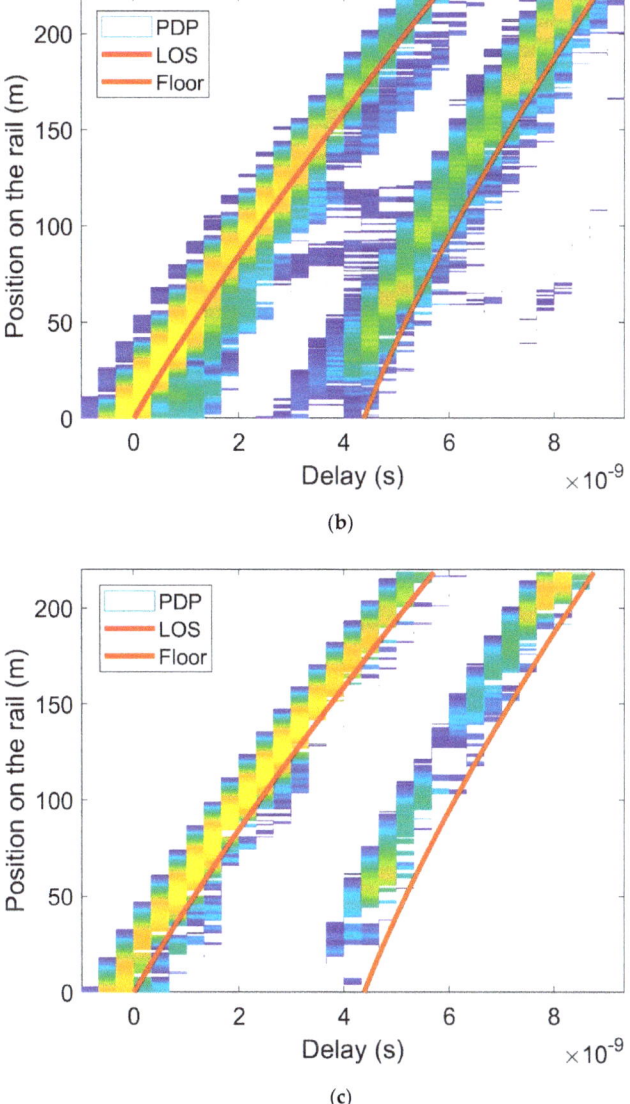

Figure 9. PDP at auditorium, at 60.5 GHz: (a) Received by antenna "2"; (b) by antenna "3"; (c) by antenna "4".

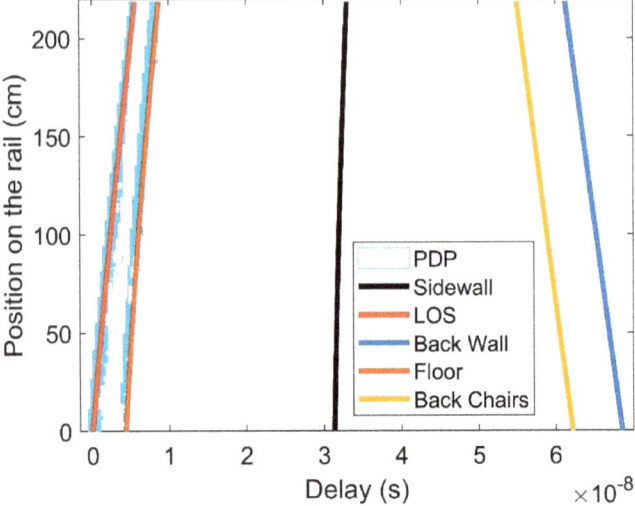

Figure 10. PDP at auditorium, at 60.5 GHz, received by antenna "2".

As stated before, time dispersion affects the maximum symbol rate that can be used without suffering ISI. In Figure 11, the RMS delay spread for all three S-parameters at both frequencies is plotted as a function of the receiver's position. The results are consistent with what was shown before (Figures 8 and 9). At 40 GHz the amount of multipath is more significant than at 60 GHz; therefore, the RMS delay spread is also larger.

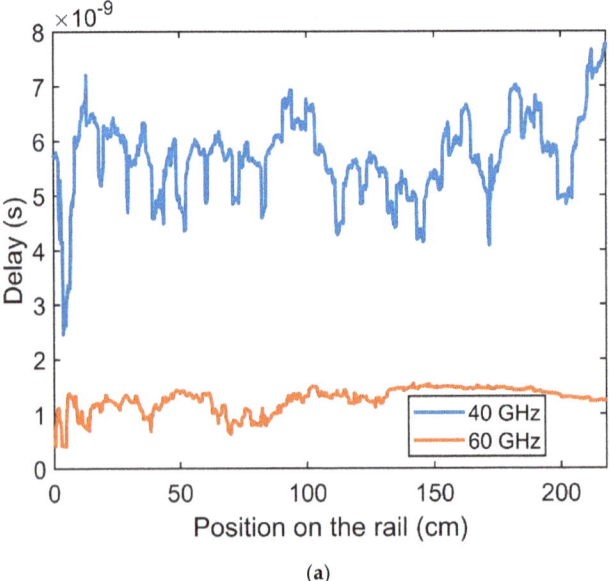

(a)

Figure 11. *Cont.*

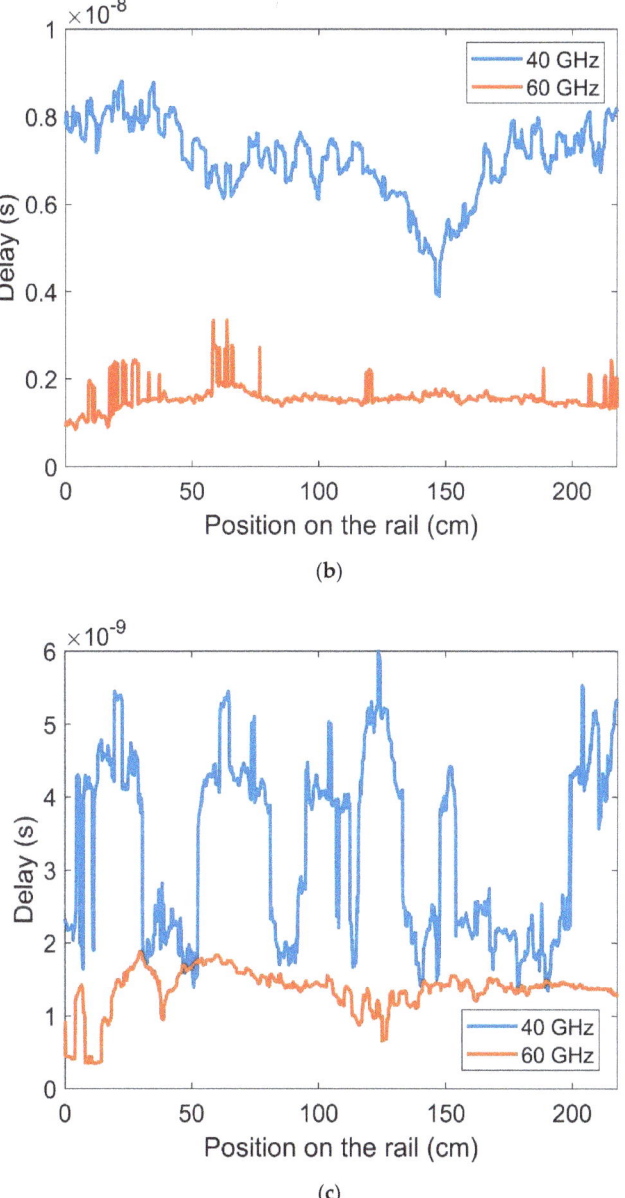

Figure 11. RMS delay at the auditorium, at both considered bands, moving the receiver along the central corridor: (**a**) Received by antenna 2; (**b**) by antenna 3; (**c**) by antenna 4.

Another interesting way to analyze the RMS delay spread values is by means of their cumulative distribution functions (CDFs). Computing the CDFs of the RMS delay spread can help to establish a reference value for the delay spread in that environment, by setting a level below which it stays for 99 percent of the time.

Taking a closer look at Table 1, it shows that the most restrictive antenna (the one with the larger value of RMS delay) is antenna "3", which is the one at the front. Hence the system must be designed to take that value into account. Again, the lower delay spread at 60 GHz, when compared to 40 GHz, is obvious in Table 1.

Table 1. RMS delay spread stays below this value (ns) 99% of the time, in the auditorium.

Band	Received by Antenna "2"	Received by Antenna "3"	Received by Antenna "4"
40 GHz	7.46	8.64	5.43
60 GHz	1.51	2.73	1.83

The second computed parameter is the coherence bandwidth. At 40 GHz, the results are very similar for all three antennas, as can be seen in Figure 12. For the signal received by antenna "2" and "4" similar behavior is observed: A level of about 800 MHz for $\alpha = 0.5$ until the antennas are moved 100 cm on the rail. The reason for this large coherence bandwidth is that at short distances the main contribution is the direct one, while the ground reflection is attenuated by the antenna pattern (as seen in Figure 9). As the distance increases, the ground reflected component is received through the center of the main lobe of the antenna pattern, and the coherence bandwidth decreases drastically. Comparing those results with the ones obtained by antenna "3", the level of the coherence bandwidth in the latter is smaller and the decay is produced at a closer distance to the transmitter.

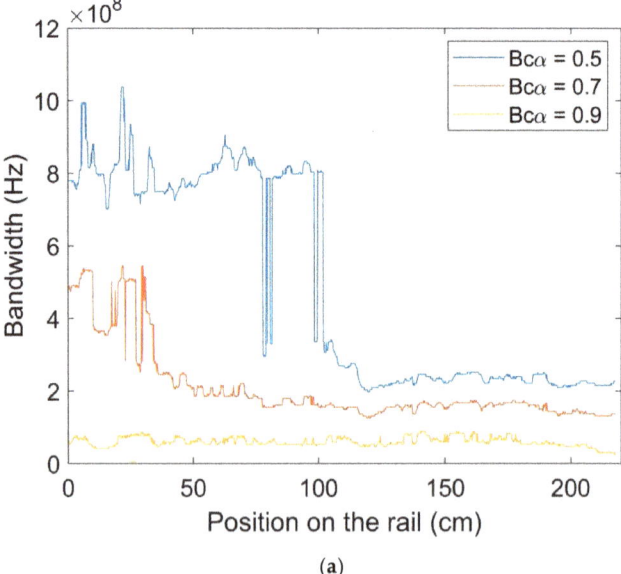

(a)

Figure 12. *Cont.*

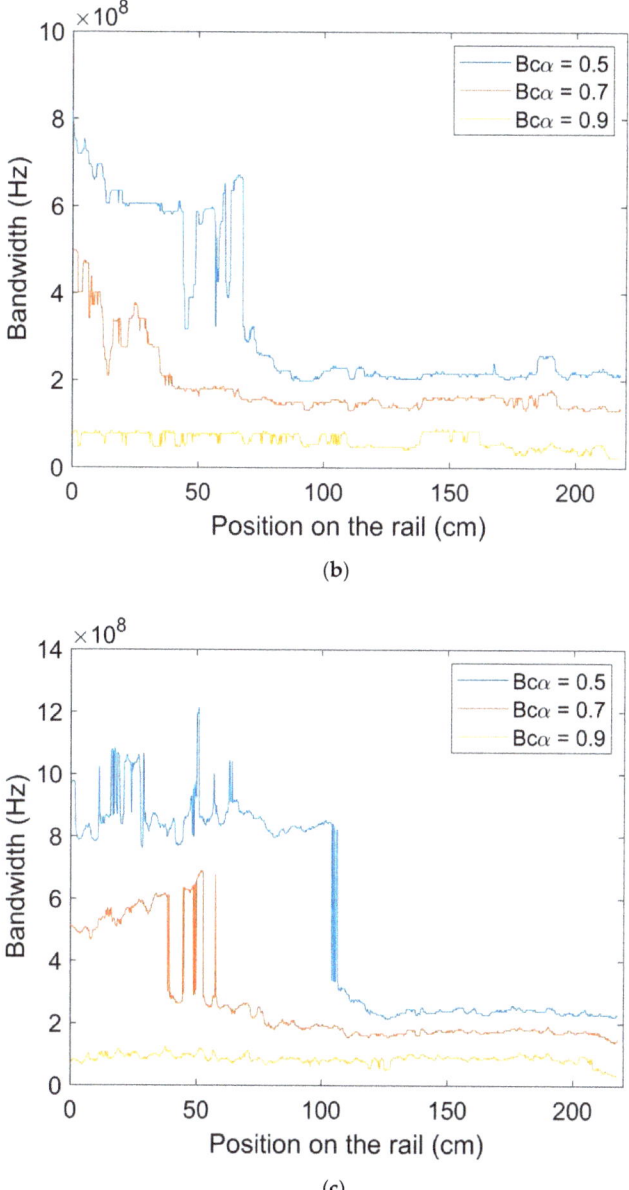

Figure 12. Coherence bandwidth as a function of the receiver position on the rail at 40 GHz received by (**a**) antenna 2; (**b**) by antenna 3; (**c**) by antenna 4.

When looking at the results for 60 GHz in Figure 13, one can observe the increase in the mean level of the coherence bandwidth for all correlation levels. This can be explained by the reduced number of echoes, compared to the 40 GHz frequency campaign.

As with the RMS delay, the percentage of the time the coherence bandwidth is above a desired threshold and is a useful metric for the design of communication systems. Here, the results for the 99th percentile can be seen in Table 2.

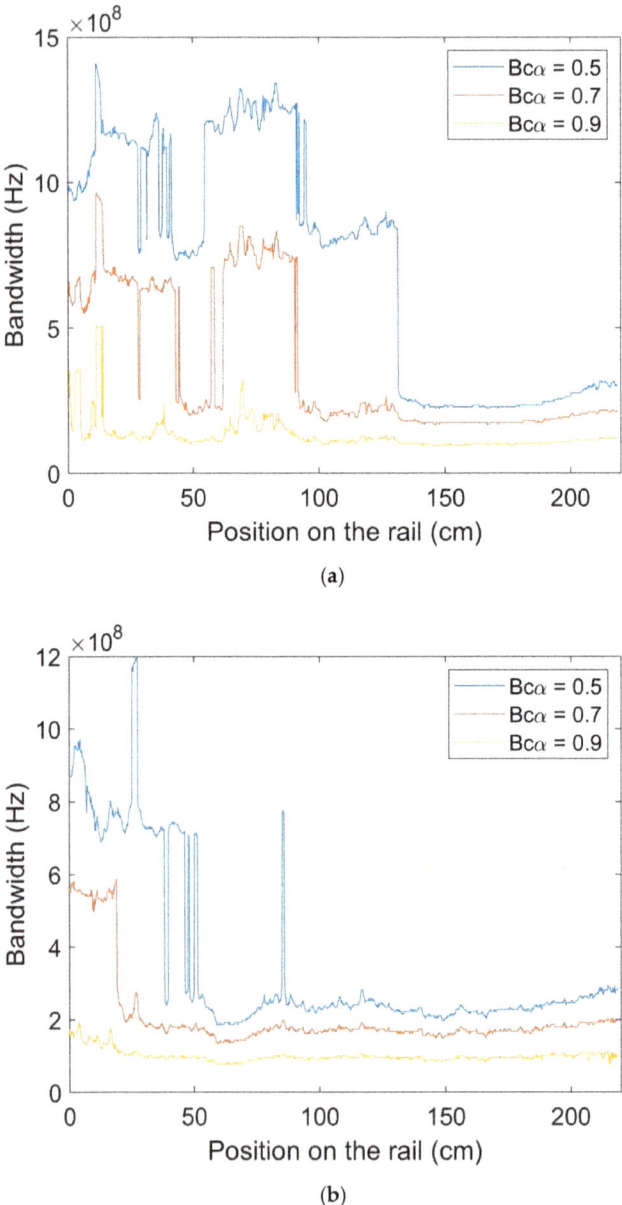

Figure 13. Cont.

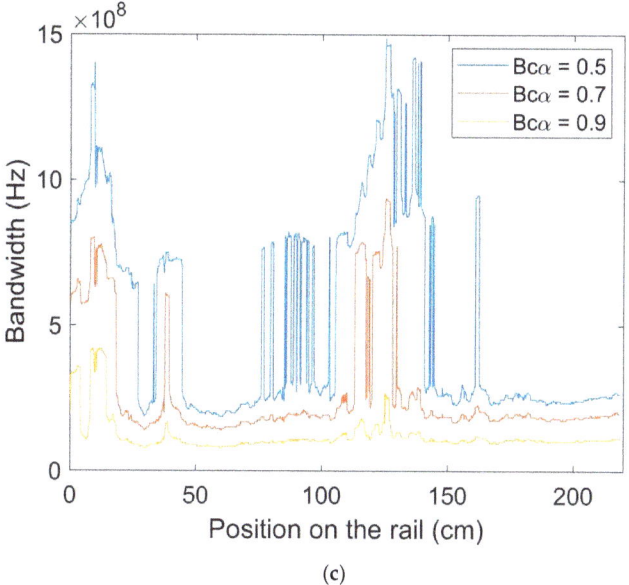

Figure 13. Coherence bandwidth as a function of the receiver position on the rail at 60 GHz received by (a) antenna 2; (b) by antenna 3; (c) by antenna 4.

Table 2. The 99th percentile of coherence bandwidth (MHz), in the auditorium.

Band	Correlation Level	Received by Antenna "2"	Received by Antenna "3"	Received by Antenna "4"
40 GHz	0.5	204	199	217
	0.7	132	132	144
	0.9	30	24	36
60 GHz	0.5	222	186	192
	0.7	168	138	144
	0.9	96	78	78

We can observe that for a coherence level of 0.9, 60 GHz clearly outperforms 40 GHz. For lower correlation levels, that behavior is not that clear. In addition, for both frequency bands, the worst scenarios are given when receiving through antenna "3".

Comparing the coherence bandwidth and the RMS delay spread values, it is clear that there is an inverse relation. The more replicas of the signal received, the larger the RMS delay spread and the lower the coherence bandwidth become. The mathematical relation is given by Fleury's limit [14,15] in Equation (5).

$$Bc \geq \frac{\cos^{-1}\alpha}{2\pi\tau_{rms}} \quad (5)$$

where Bc is the coherence bandwidth for a correlation level α and τ_{rms} is the RMS delay spread.

Figure 14 illustrates both theoretical lower limits for the coherence bandwidth with different correlations and the results obtained by analyzing the available data and plotting them as a function of RMS delay at 60 GHz.

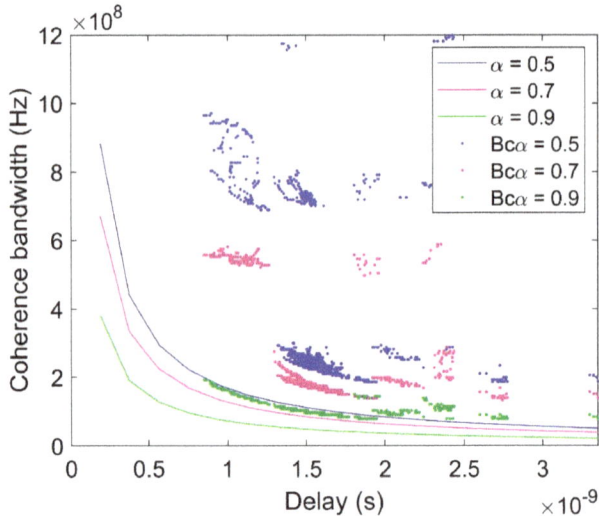

Figure 14. Theoretical Fleury limit and measured results, for different correlation levels α = 0.5, 0.7, or 0.9, at the auditorium in the 60 GHz band, received by antenna "3".

3.2. Corridor

Wideband results obtained in the corridor, in both LOS and OLOS, were processed in a similar way. Tables 3 and 4 provide RMS delay spread values and Tables 5 and 6 coherence bandwidth values, for LOS and OLOS conditions.

Table 3. RMS delay stays below this value (ns) 99% of the time, in the corridor, LOS.

Band	Channel 21	Channel 31	Channel 41
40 GHz	7.4	8.4	5.4
60 GHz	0.1	0.1	0.1

Table 4. RMS delay stays below this value (ns) 99% of the time, in the corridor, OLOS.

Band	Channel 21	Channel 31	Channel 41
40 GHz	6.9	8.4	8.7
60 GHz	4.1	5.1	3.7

Observing Tables 3 and 4, one can notice a clear difference between 40 GHz and 60 GHz. Both in LOS and OLOS, the 60 GHz signals demonstrate a lower delay spread. Indeed, the lower the frequency the more multipath is observed in the measurement campaign. Obviously, the multipath scheme is the same in both environments. However, the attenuation at 60 GHz is stronger than at 40 GHz as many multipath components at 60 GHz arrive at the receiver with a level below its sensitivity, and hence do not contribute to the received signal. Less multipath components result in a lower RMS delay spread.

The difference in RMS delay spread between LOS and OLOS is clear at 60 GHz, with lower values in LOS and with a stronger dominant direct path than in OLOS. The difference is not so clear at 40 GHz.

In general, the coherence bandwidth in these corridor scenarios is higher in the first part of the rail. After a certain distance, it drops to nearly half its value. The drop appears around 130 cm for 60 GHz and 150 cm for 40 GHz. There are two explanations for this behavior. The first one is that the antenna quits the main lobe, resulting in a weaker value in the PDP. The second one is that the multipath components are reduced, due to the environment geometry.

In the LOS scenario, the coherence bandwidth is higher at 40 GHz than at 60 GHz, as can be seen in Table 5. As previously stated, this is expected due to the higher amount of multipath. In OLOS (Table 6) the 60 GHz seems to be higher, but this is due to the absence of received signal at some places along the path.

Table 5. The 99th percentile of coherence bandwidth (MHz), in the corridor, LOS.

Band	Correlation Level	Received by Antenna "2"	Received by Antenna "3"	Received by Antenna "4"
40 GHz	0.5	282	264	306
	0.7	144	156	191
	0.9	12	12	12
60 GHz	0.5	209	179	227
	0.7	48	18	12
	0.9	12	6	6

Table 6. The 99th percentile of coherence bandwidth (MHz), in the corridor, OLOS.

Band	Correlation Level	Received by Antenna "2"	Received by Antenna "3"	Received by Antenna "4"
40 GHz	0.5	96	84	78
	0.7	72	66	60
	0.9	4	4	4
60 GHz	0.5	126	108	120
	0.7	96	84	90
	0.9	54	48	54

Regarding the theoretical limits for the coherence bandwidth, Figure 15 shows the Fleury's limit for the case of the measurements in the 40 GHz band at the corridor.

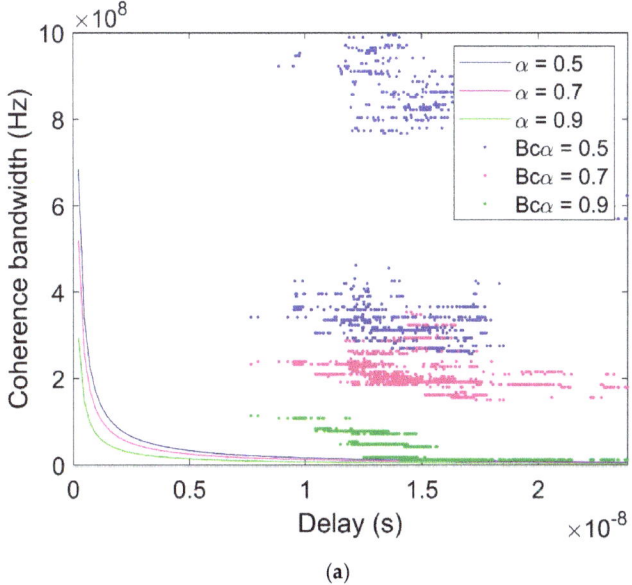

(a)

Figure 15. *Cont.*

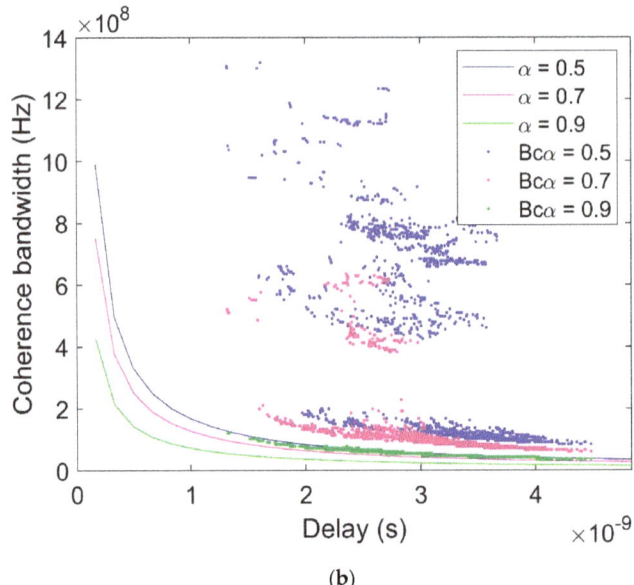

Figure 15. Theoretical Fleury's limit and measured results, for different correlation levels $\alpha = 0.5, 0.7$, or 0.9 at the corridor in the 40 GHz band: (**a**) LOS conditions, (**b**) OLOS conditions.

4. Discussion

Throughout the paper, several results provide insight into the behavior of the indoor radio channel at different environments in two clearly different millimeter-wave frequency bands: The licensed one near 40 GHz and the unlicensed around 60 GHz. Their characteristics allow the definition of advantages and disadvantages of both for deploying high rate wireless networks, which is the aim of this work.

From the measurement data and the processing of this gathered information, the main observed differences are:

1. The amount of multipath contributions is clearly reduced in 60 GHz compared to 40 GHz. This is due to the larger propagation attenuation and lower reflection coefficients at 60 GHz. At 60 GHz the energy will be scattered in a more diffuse way as the wavelength is smaller compared to the surface irregularities.
2. This leads to smaller delay spreads at 60 GHz than at 40 GHz (i.e., 2.73 ns versus 8.64 ns at the auditorium).
3. The radio channel is more frequency selective at 40 GHz, with coherence bandwidths for a 0.9 correlation level of 24 MHz at 40 GHz in the auditorium, while a value of 78 MHz was measured in the same environment at 60 GHz.
4. Regarding the different scenarios in LOS conditions, smaller environment dimensions yield to a smaller RMS delay spread at 60 GHz. A value of 0.1 ns is obtained in the corridor while 2.73 ns is obtained in the auditorium. At 40 GHz the difference is not so clear (8.4 ns versus 8.64 ns).
5. On the other hand, smaller scenarios correspond to larger coherence bandwidth values, particularly at 60 GHz, with coherence bandwidth values for a 0.9 correlation level increasing from 6 MHz in the corridor to 78 MHz in the auditorium.
6. OLOS conditions result in larger delay spreads (5.1 ns) compared to LOS conditions (0.1 ns) at 60 GHz. The difference at 40 GHz is no so clear (8.7 ns versus 8.4 ns).

7. The inverse relation between the delay spread and coherence bandwidth has been confirmed by plotting both parameters against Fleury's limit.
8. Due to the reduced multipath propagation conditions at 60 GHz, channel parameters seem to be more sensitive to changes in the size of the environment and on the visibility conditions (LOS/OLOS) than at 40 GHz.
9. Regarding ray-tracing simulations, a simple ray-tracing model with just single reflections would be enough to represent all relevant multipath contributions at 60 GHz. Higher order reflections should be considered for modelling all contributions at 40 GHz.

At 38 GHz, results for the corridor agree with those in the literature [16] where measurements were performed in a wagon with similar dimensions and geometry. The delay spread values reported are 7.63 ns and 9.77 ns, very similar to the 8.4 ns we measured in LOS condition.

At 60 GHz, similar delay spread values can be found in [17] where a mean delay spread value of 6 ns with 2 ns standard deviation are reported for a conference room and in [18] where delay spreads below 7 ns are measured for 99% of the positions, also at a similar conference room. Finally, in [19] a value of 3.13 ns is found, but for a smaller environment.

5. Conclusions

A wideband measurement campaign at two millimeter-wave frequency bands, 40 GHz and 60 GHz, has been conducted in two indoor scenarios: An auditorium and a narrow corridor. Measurements included both LOS and OLOS situations.

The results have been processed and two wideband channel parameters have been obtained: The RMS delay spread and the coherence bandwidth. Based on these two parameters, the performance of both frequency bands is compared.

We conclude that there are more multipath contributions at 40 GHz than at 60 GHz, resulting in a larger delay spread and larger frequency selectivity at 40 GHz than at 60 GHz. The radio channel at 60 GHz seems to be more sensitive to any change in the size of the environment or in the visibility conditions (LOS/OLOS).

We have also concluded that multipath contributions at 60 GHz can be explained by single reflections on the environment scatterers, while higher order reflections should be considered to model all multipath contributions at 40 GHz.

The results in this paper should be taken into account if a decision has to be made on the frequency band to be used in an indoor link for a wireless communication system. Results also provide a deep insight into the physical propagation mechanisms that take place and help to understand how these mechanisms determine the differences in the wideband parameters of the radio channel. Finally, results will be valuable also for those developing ray-tracing models for the indoor radio channel at millimeter-wave frequencies.

Author Contributions: Design of the measurement campaign M.G.S. and I.C.; measurement system setup, measurements and data processing: R.H. and M.R.; analysis of results: M.G.S., I.C. and J.V.; writing: all.

Funding: This research was funded by the Spanish Government, Ministerio de Ciencia, Innovación y Universidades, Secretaría de Estado de Investigación, Desarrollo e Innovación, grant number TEC2017-85529-C3-3R); Xunta de Galicia, grant number ED431C 2019/26; atlanTTic Research Center; and the European Regional Development Fund (ERDF).

Acknowledgments: The authors would like to thank Isabel Expósito for her technical support in the management of the radioelectric equipment.

Conflicts of Interest: The authors declare no conflict of interest.

References

1. Marcus, M.J. 5G and "IMT for 2020 and beyond" [Spectrum Policy and Regulatory Issues]. *IEEE Wirel. Commun.* **2015**, *22*, 2–3. [CrossRef]

2. Rappaport, T.S.; Sun, S.; Mayzus, R.; Zhao, H.; Azar, Y.; Wang, K.; Wong, G.N.; Schulz, J.K.; Samimi, M.; Gutierrez, F. Millimeter Wave Mobile Communications for 5G Cellular: It Will Work! *IEEE Access* **2013**, *1*, 335–349. [CrossRef]
3. European 5G Observatory "US 5G Auction, in the 24 GHz Band". Available online: http://5gobservatory.eu/us-5g-auction-in-the-24-ghz-band/ (accessed on 21 September 2019).
4. ITU-R Recommendation M.2003-1. *Multiple Gigabit Wireless Systems in Frequencies around 60 GHz*; Recommendation: Geneva, Switzerland, 2015.
5. Pahlavan, K.; Howard, S.J. Frequency domain measurements of indoor radio channels. *Electron. Lett.* **1989**, *25*, 1645–1647. [CrossRef]
6. Varela, M.S.; Sánchez, M.G. An improved method to process measured radio-channel impulse responses. *Microw. Opt. Technol. Lett.* **2000**, *24*, 158–162. [CrossRef]
7. Ai, B.; Guan, K.; He, R.-S.; Li, J.-Z.; Li, G.-K.; He, D.; Zhong, Z.-D.; Huq, K.M.S. On Indoor Millimeter Wave Massive MIMO Channels: Measurement and Simulation. *IEEE J. Sel. Areas Commun.* **2017**, *35*, 1678–1690. [CrossRef]
8. Cuiñas, I.; Sanchez, M.G. Measuring, modeling, and characterizing of indoor radio channel at 5.8 GHz. *IEEE Trans. Veh. Technol.* **2001**, *50*, 526–535. [CrossRef]
9. Zahedi, Y.; Ngah, R.; Nunoo, S.; Mokayef, M.; Alavi, S.; Amiri, I.S. Experimental measurement and statistical analysis of the RMS delay spread in time-varying ultra-wideband communication channel. *Measurement* **2016**, *89*, 179–188. [CrossRef]
10. Rappaport, T.S. *Wireless Communications: Principles and Practice*; Prentice Hall: Piscataway, NJ, USA, 1996.
11. He, Y.; Yin, X.; Ji, Y.; Lu, S.; Du, M. Spatial characterization of coherence bandwidth for 72 GHz mm-wave indoor propagation channel. In Proceedings of the 2015 9th European Conference on Antennas and Propagation (EuCAP), Lisbon, Portugal, 13–17 April 2015; pp. 1–5.
12. Ghaddar, M.; Talbi, L.; Delisle, G. Coherence bandwidth measurement in indoor broadband propagation channel at unlicensed 60 GHz band. *Electron. Lett.* **2012**, *48*, 795. [CrossRef]
13. Balanis, C.A. *Advanced Engineering Electromagnetics*; John Wiley & Sons: New York, NY, USA, 1989.
14. Fleury, B.H. An uncertainty relation for WSS processes and its application to WSSUS systems. *IEEE Trans. Commun.* **1996**, *44*, 1632–1634. [CrossRef]
15. Alejos, A.V.; Sánchez, M.G.; Cuiñas, I.; Dawood, M. Wideband noise radar based in phase coded sequences. In *Radar Technology*; InTech: Rijeka, Croatia, 2009; pp. 39–60. ISBN 978-953-307-029-2.
16. Rubio-Arjona, L.; Rodrigo-Peñarrocha, V.M.; Molina-Garcéa-Pardo, J.-M.; Juan-Llácer, L.; Pascual-García, J. Millimeter wave channel measurements in an intra-wagon environment. *IEEE Trans. Veh. Technol.* **2019**. [CrossRef]
17. Semkin, V.; Karttunen, A.; Jarvelainen, J.; Andreev, S.; Koucheryavy, Y. Static and Dynamic Millimeter-Wave Channel Measurements at 60 GHz. In Proceedings of the 12th European Conference on Antennas and Propagation (EuCAP 2018), London, UK, 9–13 April 2018; pp. 1–5.
18. Bamba, A.; Mani, F.; D'Errico, R. Millimeter-Wave Indoor Channel Characteristics in *V* and *E* Bands. *IEEE Trans. Antennas Propag.* **2018**, *66*, 5409–5424. [CrossRef]
19. Martinez-Ingles, M.-T.; Pascual-García, J.; Rodríguez, J.-V.; Molina-Garcia-Pardo, J.-M.; Juan-Llácer, L.; Gaillot, D.P.; Liénard, M.; Degauque, P. Indoor radio channel characterization at 60 GHz. In Proceedings of the 2013 7th European Conference on Antennas and Propagation (EuCAP), Gothenburg, Sweden, 8–12 April 2013; pp. 2796–2799.

© 2019 by the authors. Licensee MDPI, Basel, Switzerland. This article is an open access article distributed under the terms and conditions of the Creative Commons Attribution (CC BY) license (http://creativecommons.org/licenses/by/4.0/).

Article

3D Printing Using a 60 GHz Millimeter Wave Segmented Parabolic Reflective Curved Antenna

Benxiao Cai [1,2,*], Lingling Sun [2] and Yuchao Lei [2]

[1] College of Electrical Engineering, Zhejiang University, Hangzhou 310027, China
[2] Key Laboratory of RF Circuits and Systems, Ministry of Education of China, Hangzhou Dianzi University, Hangzhou 310018, China; sunll@hdu.edu.cn (L.S.); albert_8@163.com (Y.L.)
* Correspondence: xiaocai@hdu.edu.cn; Tel.: +86-571-86919037

Received: 16 January 2019; Accepted: 29 January 2019; Published: 11 February 2019

Abstract: This paper proposes a segmented parabolic curved antenna, which can be used in the base station of a 60 GHz millimeter wave communication system, with an oblique Yagi antenna as a feed. By analyzing the reflection and multi-path interference cancellation phenomenon when the main lobe of the Yagi antenna is reflected, the problem of main lobe splitting is solved. 3D printing technology relying on PLA (polylactic acid) granule raw materials was used to make the coaxial connector bracket and segmented parabolic surface. The reflective surface was vacuum coated (via aluminum evaporation) with low-loss aluminum. The manufacturing method is environmentally friendly and the structure was printed with 0.1 mm accuracy based on large-scale commercial applications at a low cost. The experimental results show that the reflector antenna proposed in this paper achieves a high gain of nearly 20 dBi in 57–64 GHz frequency band and ensures that the main lobe does not split.

Keywords: Yagi antenna; parabolic; 3-D printing; millimeter wave

---

1. Introduction

With the rapid development of broadband technology, there is a large potential market demand for wireless communication with 1 Gbit capacity per second. This has led to 60 GHz millimeter wave communication becoming a research hotspot. High gain, a wide bandwidth, no main-lobe splitting, easy integration and low loss are key indicators for 60 GHz millimeter wave base station antennas. On the other hand, the silicon-based system-on-chip system concentrates the on-chip antenna and the on-chip active modules in the same chip, which is advantageous for mass production. However, due to the loss of the silicon substrate, the conventional on-chip antenna cannot meet the requirements to be a base station antenna. The substrate integrated antenna [1–4] and the microstrip antenna [5] can meet the above requirements but to achieve high gain, the above two antennas need to adopt an array combination and the feed network is very complicated. This also increases the profile of the antenna, which means more corrosive liquid in manufacturing, causing consequent environmental pollution problems. Therefore, the use of an on-chip reflective antenna is alternative choice. In the literature [6], multiple planar dipole antennas are used as feeds and concentric circles of different radii are formed by hollowing out the layers of the multilayer substrate. The cross section is approximately parabolic and has a directional beam scanning function but its bandwidth is only 4 GHz, which is unable to meet the requirements of millimeter wave broadband communication. In contrast, PLA plastics used in plastic powder additive manufacturing technology (i.e., 3D printing technology) are naturally degradable and environmentally friendly, while substrates made by 3D printing can be formed by vacuum plating to form reflective surfaces for antennas. The 3D printing accuracy of existing commercial applications can reach a range of 0.1 mm; with low cost, which is very suitable for the millimeter wave frequency band. Compared with the traditional metal reflective surface, the plastic printing plus technology vacuum plating can reduce the weight by 80% and its structure is light. While studies reflected in the literature [7–9] made a useful attempt, the feed antenna used for a waveguide interface and cannot be seamlessly integrated with active modules. In view of the above problems, this paper adopts 3D printing technology to realize the antenna reflection surface and utilize the planar Yagi antenna as the feed source, hence achieving seamless integration of the on-chip antenna and the active module. Moreover, in order to meet the key indicators of the above 60 GHz millimeter wave base station antenna, this paper adopts the oblique reflection surface structure, which eliminates the shadowing effect of the feed and eliminates the main lobe splitting caused by multipath interference cancellation. The performance of the antenna is improved effectively. Through optimization, the antenna proposed in this paper satisfies the above indicators during actual measurement and is expected to become a new generation of on-chip integrated antennas for base stations.

2. Antenna Design

2.1. Parabolic Feed Antenna Design

The Yagi-Uda antenna [10] itself is composed of end-fire arrays with high gain, wide bandwidth and low cross-polarization. The planar Yagi-Uda antenna is extremely suitable for microwave and millimeter wave applications due to its high gain, low loss, high radiation efficiency and ease of fabrication [10]. Therefore, it was chosen as the feed for the reflector antenna in this paper. The planar Yagi-Uda antenna includes the extended ground wire on the PCB as a reflector and a planar printed dipole as the driver, both are printed on both sides of the PCB. All the four directors of Yagi antenna were printed paralleled with the dipole on the upper surface of PCB. The initial design was based on empirical data from free space and then scaled down to the equivalent dielectric constant. The microstrip, the dipole and the directors were connected by planar printing balun to ensure sufficient bandwidth. However, this design also created a high level of cross-polarization. Therefore, the electro-magnetic field numerical analysis method (HFSS) was used for analysis and adjustment. The method was used to increase the shape of the saw ruler to eliminate surface waves and reduce cross-polarization. As shown in reference [10], the designed Yagi antenna has four directors. The substrate is Rogers 5880 (dielectric constant of 2.2) and the substrate thickness 0.254 mm. Figure 1 shows the connector and connector bracket of the antenna.

Figure 1. Connector Bracket and connector.

The Yagi antenna has a −10 dB bandwidth of 56–67 GHz which is shown in Figure 2. It can demonstrate a low sidelobe radiation pattern with a maximum gain of 9.3 dBi [10]. It was used as a parabolic antenna feed, supplemented by a 3D printed support frame with an accuracy of 0.1 mm, hence achieving a seamless connection of the transfer interface. In the 60 GHz millimeter wave environment, the tilt of the electrical scale by 1 mm to 2 mm will bring a large error. Therefore, the support bracket was designed to avoid the tilt of the connector due to gravity and the system accuracy was improved.

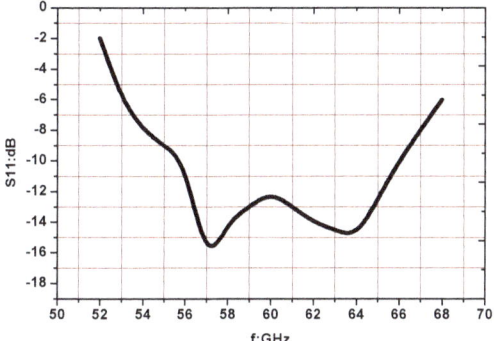

Figure 2. S parameter of Yagi-Uda antenna.

2.2. Parabolic Antenna Structure

The parabolic antenna shown in Figure 3a affects the antenna performance by obscuring the reflected parallel electro-magnetic waves due to the size of the feed at the parabolic focus. The cylindrical paraboloid shown in Figure 3b was used to allow the antenna feed to be incident obliquely and the reflected parallel electromagnetic waves do not create a shadowing effect. In order to obtain the parabolic profile of the reflecting cylinder, the paraboloid profile is satisfied as follows:

$$\rho = \frac{2p}{1+\cos\phi} = p \cdot \sec^2\left(\frac{\phi}{2}\right) \quad (1)$$

where ρ represents the distance from the center of the parabola to the curve in polar coordinates, p is double the focal distance F as shown in Figure 3 and ϕ is the angle in polar coordinates.

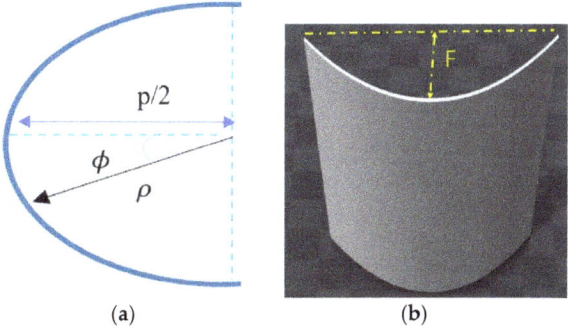

Figure 3. Parabolic reflector. (**a**) Parabolic curve (**b**) 3D structure.

In order to make the antenna sidelobe smaller, the angle α between the reflected wave and the incident wave was 120° as shown in Figure 4a after optimization. The focal length F of the parabolic was 7 mm and p was 14 mm as shown in Figure 3a. The parabolic profile has a profile radius R of 20 mm. In the optimization process, we choose the angle α between the reflected wave and the incident wave to be 100°, 120° and 140°, respectively. The radiation patterns in Figure 5 are obtained by HFSS simulation.

The parabolic cross section is shown in Figure 3a and the cylindrical parabolic antenna is shown in Figure 6a. The Yagi antenna is placed horizontally and perpendicular to the support column and the angle between the paraboloid and the support column is 60° according to α equal to 120°. The paraboloid was cut with a spherical shape to make the contour more regular.

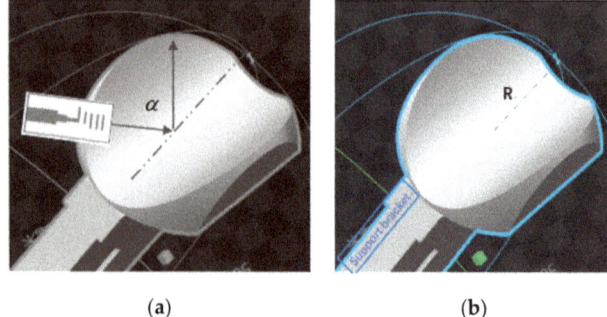

Figure 4. Cylindrical parabolic antenna (**a**) Angle α (**b**) Profile radius R.

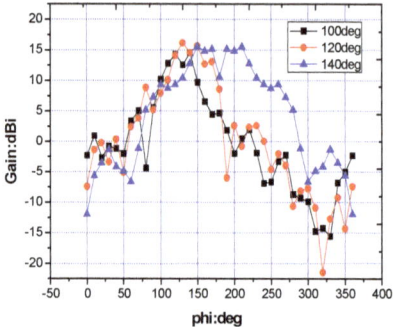

Figure 5. Radiation patterns with different angle α measurements at 59 GHz.

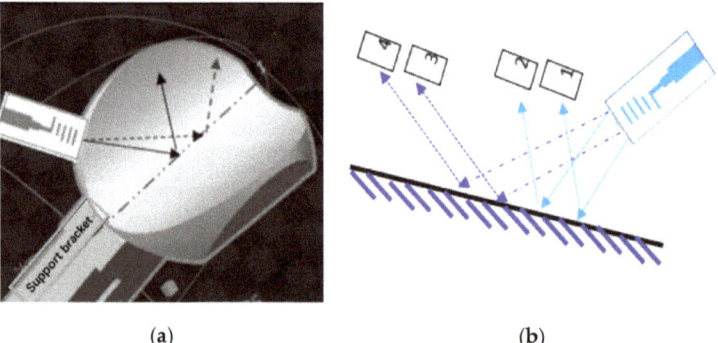

Figure 6. Optical paths in the parabolic reflector (**a**) 3D diagram (**b**) 2D diagram.

The main lobe of the Yagi antenna has a certain narrow −3 dB width. The optical path of the longitudinal section is shown in Figure 6. As illustrated in Figure 6b, the optical path difference between optical paths 1 and optical path 2 is much shorter than that of optical path 3 and optical path 4. Therefore, there is always one direction and the optical path difference of the two optical paths is half the wavelength, which causes the main lobe to be split due to interference. Both the simulation and measurement result of the main lobe splitting are shown in Figure 7. The measurement splitting is more obvious and while the simulation splitting has a 2 dB recess.

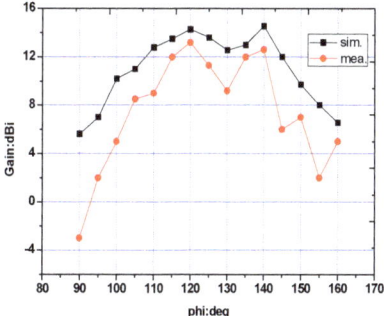

Figure 7. Simulation and test data for the cylindrical reflector antenna at 59 GHz.

2.3. Parabolic Curved Antenna

In order to solve the afore-mentioned main lobe splitting problem, the reference [11] used the reflective surface of the longitudinal parabola profile. As shown in Figure 8a, at this time, the reflected rays are almost parallel and the interference phenomenon can be effectively avoided. The paraboloid was also modified from the original cylinder, as shown in Figure 6a, to an orthogonal parabolic surface, as shown in Figure 8b. The distance between the four vertices of the paraboloid and the center of the cutting sphere is R and R is contour radius. Moreover, the distance between the top and bottom vertices on the same side is 1.8 R.

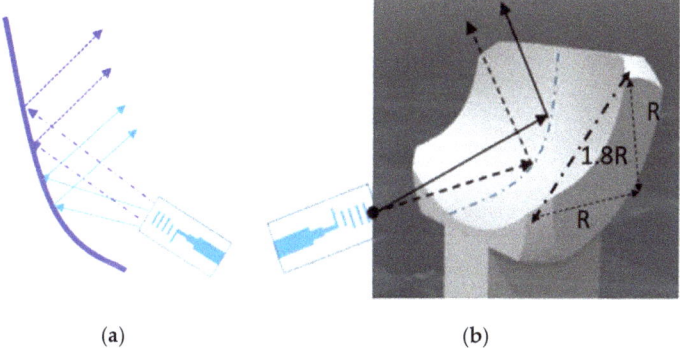

Figure 8. Rays on the longitudinal parabola (a) 2D diagram (b) 3D diagram.

By calculating the formula based on the parabolic antenna gain:

$$G = \eta_A 4\pi S / \lambda^2 \qquad (2)$$

Here η_A is the aperture efficiency of the parabolic antenna and it includes spillover efficiency η_s and aperture taper efficiency η_t. S is the physical area of the antenna aperture and λ is the wavelength. As the flare angle increases with contour radius R, η_s increases correspondingly. Since here the gain value decreases with R as shown in Figure 9 (when R = 20, 30, 40, the gain pattern can be referred to in Figure 10), it indicates that the irradiation becomes non-uniform due to decrease of the aperture taper efficiency η_t. When the contour radius is too small, the side lobe level become higher due to the diffraction effect of the edge or small spillover efficiency η_s. The highest side-lobe according to Figure 9 is 7 dBi, corresponding to a maximum main lobe gain of 20 dBi. Regarding optimization, the optimum profile radius was 32 mm (3.7 dBi for the side lobes and 20 dBi for the main lobes). Considering that the feed will deviate from the vertical direction of support column due to alignment error, the minimum distance between the contour edge and the projection point of the feed center on the paraboloid was guaranteed to be 32mm and the contour radius was set at 40mm. Simultaneously, the maximum gain is 19.3 dBi at 61 GHz frequency.

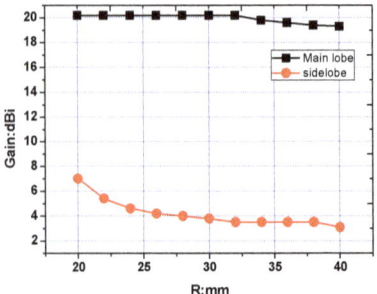

Figure 9. Gain values of the main and sidelobes at different contour radii. at 59 GHz.

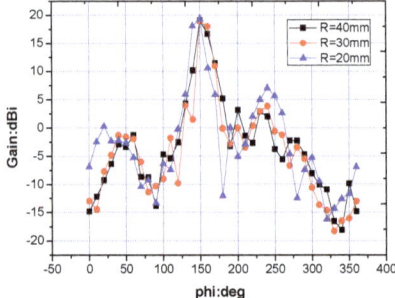

Figure 10. Gain pattern with different parabolic surface radii. at 59 GHz.

2.4. Segmented Parabolic Surface Antenna

The parabola converted the ideal point source at the focus into parallel rays. However, for a given −3 dB beam width feed like the Yagi antenna, it was viewed as a collection of multipoint sources at different locations. Therefore, the parabola needs to be corrected in order to integrate the point sources at different locations to obtain the optimal parallel light generation effect. The parabolic equation is:

$$\rho = \frac{2p}{1+\cos\phi} \qquad (3)$$

A segmentation combination of the parabolic equation was made using different parameter measures for p, which is similar to what was done in reference [12]. Optimized using the HFSS simulation, the combined parabolic equation is:

$$\begin{cases} \rho = \frac{2p_1}{1+\cos\phi}; & \phi \in [-\frac{\pi}{6}, \frac{\pi}{6}] \\ \rho = \frac{2p_2}{1+\cos\phi}; & \phi \in [-\frac{7\pi}{36}, -\frac{\pi}{6}] \cup [\frac{\pi}{6}, \frac{7\pi}{36}] \\ \rho = \frac{2p_3}{1+\cos\phi}; & \phi \in [-\frac{\pi}{4}, -\frac{7\pi}{36}] \cup [\frac{7\pi}{36}, \frac{\pi}{4}] \end{cases} \qquad (4)$$

The value of p_1, p_2, p_3 are 12, 14 and 16 mm respectively. At 61 GHz, the gain of the segmented combined parabolic surface was 1 dB higher than the gain of the parabolic surface, with a gain of 20.5 dBi. As can be seen from Figure 11, the gain value of each frequency point at 57–64 GHz increased stably by 1 dB.

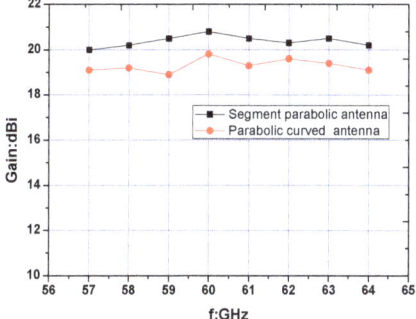

Figure 11. Comparison between Segment parabolic and Parabolic curved antenna.

3. Simulation and Experimental Results

The three-dimensional structure of the segmented parabolic surface is shown in the Figure 12 above which Figure 12c shows the antenna in the microwave anechoic chamber, and its one-dimensional graph is shown in Figure 13. In Figure 14, two curves represent the simulation result and the measured result respectively. At 61 GHz, the measured gain was 20 dBi and the first side lobes reached 6 dBi, which is 14 dB lower than the main lobe. At 64 GHz, the first side lobes reached 7 dBi, which is 13 dB lower than the main lobe.

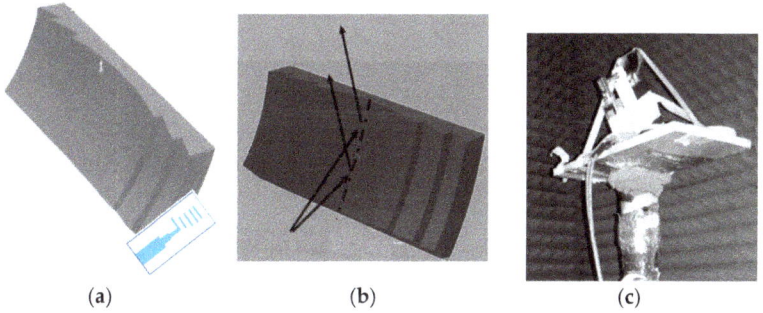

Figure 12. (a) Three-dimensional structure of the segmented parabolic surface (b) Rays on segmented parabolic surface (c) The segmented parabolic surface in the microwave anechoic chamber.

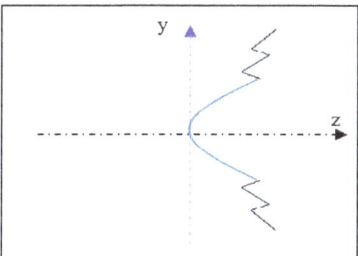

Figure 13. Segment parabolic curve antenna.

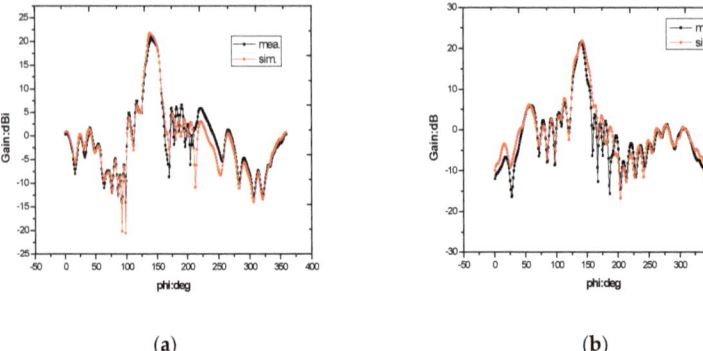

Figure 14. Simulation and test gain patterns at frequencies of (a) 61 and (b) 64 GHz.

4. Conclusions

An antenna system using a 3D printed vacuum-plated parabolic surface and a planar Yagi antenna as a feed can be seamlessly integrated as planar antenna with on-chip millimeter active module. The improved reflective surface design as shown in Figure 8, by using a parabolic surface, gain was not only improved by 5 dB but the gain value reached 19 dBi. Also, it improved the main-lobe splitting as shown in Figure 7. Moreover, the segmented combined parabolic surface as shown in Figure 12 was further improved and the gain value reached 20 dBi in the 57–64 GHz band. Compared with the traditional metal reflective surface, the plastic printing plus vacuum plating parabolic can reduce the weight by 80% and it is easy fabricated, low cost and environment friendly.

Author Contributions: B.C. completed the methodology; B.C. and Y.L. completed the design; B.C. wrote the paper. L.S. reviewed the paper.

Funding: This research was funded by the Natural Science Foundation of Zhejiang Provincial education department under Grant Y201636542.

Acknowledgments: The authors would like to thank Lei Liu for his supports in the simulations.

Conflicts of Interest: The authors declare no conflict of interest.

References

1. Yang, T.Y.; Wei, H.; Zhang, Y. Wideband millimeter-wave substrate integrated waveguide cavity-backed rectangular patch antenna. *IEEE Antennas Wirel. Propag. Lett.* **2014**, *13*, 205–208.
2. Li, M.J.; Luk, K.M. A low-profile unidirectional printed antenna for millimeter-wave applications. *IEEE Trans. Antennas Propag.* **2013**, *62*, 1232–1237.
3. Cheng, S.; Yousef, H.; Kratz, H. 79 GHz slot antennas based on substrate integrated waveguides (SIW) in a flexible printed circuit board. *IEEE Trans. Antennas Propag.* **2009**, *57*, 64–71.
4. Guntupalli, A.B.; Wu, K. 45° linearly polarized high-gain antenna array for 60-GHz radio. *IEEE Antennas Wirel. Propag. Lett.* **2014**, *13*, 384–387.
5. Lamminen, A.E.I.; Saily, J.; Vimpari, A.R. 60-GHz patch antennas and arrays on LTCC with embedded-cavity substrates. *IEEE Trans Antennas Propag.* **2008**, *56*, 2865–2874.
6. Hosseini, A.; Kabiri, S.; De Flaviis, F. V-Band High-Gain Printed Quasi-Parabolic Reflector Antenna with Beam-Steering. *IEEE Trans. Antennas Propag.* **2017**, *65*, 589–591.
7. D'Auria, M.; Otter, W.J.; Hazell, J.; Gillatt, B.T.; Long-Collins, C.; Ridler, N.M.; Lucyszyn, S. 3-D printed metal-pipe rectangular waveguides. *IEEE Trans. Compon. Packag. Manuf. Technol.* **2015**, *5*, 1339–1349.
8. Nguyen, N.T.; Delhote, N.; Ettorre, M.; Baillargeat, D.; Coq, L.; Sauleau, R. Design and characterization of 60-GHz integrated lens antennas fabricated through ceramic stereolithography. *IEEE Trans. Antennas Propag.* **2010**, *58*, 2757–2762.

9. Nayeri, P.; Liang, M.; Sabory-Garcı, R.A.; Tuo, M.; Yang, F.; Gehm, M.; Xin, H.; Elsherbeni, A.Z. 3D printed dielectric reflectarrays: Low-cost highgain antennas at sub-millimeter waves. *IEEE Trans. Antennas Propag.* **2014**, *62*, 2000–2008.
10. Alhalabi, R.A.; Chiou, Y.C.; Rebeiz, G.M. Self-shielded high-efficiency Yagi-Uda antennas for 60 GHz communications. *IEEE Trans. Antennas Propag.* **2011**, *59*, 742–750.
11. Xie, Z.M.; Wang, L.X.; Pei, S.S. Linear array feeding caustics parabolic curved antenna for space power synthesis. *Chin. J. Radio Sci.* **2012**, *2*, 209–214.
12. Rao, S.; Tang, M.Q. Stepped-reflector antenna for dual-band multiple beam satellite communications payloads. *IEEE Trans. Antennas Propag.* **2006**, *54*, 801–811.

© 2019 by the authors. Licensee MDPI, Basel, Switzerland. This article is an open access article distributed under the terms and conditions of the Creative Commons Attribution (CC BY) license (http://creativecommons.org/licenses/by/4.0/).

Article

Beam Scanning Capabilities of a 3D-Printed Perforated Dielectric Transmitarray

Andrea Massaccesi *, Gianluca Dassano and Paola Pirinoli

Department of Electronics and Telecommunications, Polytechnic University of Turin, 10129 Turin, Italy; gianluca.dassano@polito.it (G.D.); paola.pirinoli@polito.it (P.P.)
* Correspondence: andrea.massaccesi@polito.it; Tel.: +39-0110904209 (ext. 4209)

Received: 11 February 2019; Accepted: 23 March 2019; Published: 28 March 2019

Abstract: In this paper, the design of a beam scanning, 3D-printed dielectric Transmitarray (TA) working in Ka-band is discussed. Thanks to the use of an innovative three-layer dielectric unit-cell that exploits tapered sections to enhance the bandwidth, a 50 × 50 elements transmitarray with improved scanning capabilities and wideband behavior has been designed and experimentally validated. The measured radiation performances over a scanning coverage of ±27° shown a variation of the gain lower than 2.9 dB and a 1-dB bandwidth in any case higher than 23%. The promising results suggest that the proposed TA technology is a valid alternative to realize a passive multibeam antenna, with the additional advantage that it can be easily manufactured using 3D-printing techniques.

Keywords: transmitarray antenna; beam scanning; planar lens; discrete lens; tapered matching; 3D-printed antenna; 3D-printing; additive manufactuing

1. Introduction

The next generation of high performing radar and satellite communication systems is requiring the development of enhanced technologies for the design and realization of its components. For what concerns antennas, the main constraints are related to the frequency bands, moving from microwave to millimeter or sub-THz frequencies, the high gain and the multibeam or beam scanning capabilities. To generate wide scanning antennas with good performances in all the pointing directions, the use of active arrays is the most straightforward solution, even if they are characterized by high complexity, that increase dramatically with the number of array elements; moreover, at the higher frequencies, the poor performances of the feeding network represent a limitation to this type of antenna. An alternative is that of using a mechanical beam-steering mechanism, easily implementable in reflector-type antennas. To reduce the antenna volume, recently, solutions where Reflectarray (RA) or Transmitarray (TA) antennas substitute the reflector are taken into account, since they represent a convenient and efficient alternative for obtaining high gain and low cost antennas [1,2]. A transmitarray is typically composed of a feed source that illuminates a single or multilayer quasi-periodic array, whose unit-cells are characterized by one or more variable geometrical parameters, through which it is possible to control the transmission coefficient phase of each cell and to obtain the desired radiation pattern [2]. The TA unit-cell can consist in metallic elements printed on different dielectric layers [3–5], in completely metallic slot-based layers separated by air gaps [6,7] or even in a dielectric structure perforated by one or several holes, as implemented in [8–13]. In comparison with reflectarrays, TAs present the advantage to not suffer for blockage, which require the use of off-set feeds in RA, and this increases the difficulty to implement efficiently the beam scanning.

In view of their features, transmitarrays have been widely considered for the realization of multibeam [14,15] or beam scanning antennas [16–27]. The most straightforward way to obtain the beam steering is that of integrating active elements in the unit-cell, such as p-i-n diodes [16–20] or varactor diodes [21–27]: the resulting radiation features are good, but the same considerations already done for active arrays apply. An alternative solution can be the use of a passive TA illuminated by a moving feed or by a feed array, generating beams that impinge on the TA surface with different incident angles and are therefore focused by the TA itself in different directions [14–18]. The main limitation of this solution is its reduced scanning capabilities: in [16], a small size configuration shows a reduction of the gain of approximately 4 dB on a scanning range from $-30°$ to $30°$. Some improvements can be obtained using a larger TA, so that just one part of it is illuminated by each incident field [15,17], or designing the TA so that it acts as a bifocal lens [18], even if in this case a degradation of the beam in the broadside direction occurs.

In this paper, the design of the 3D-printable dielectric transmitarray with beam-scanning capabilities is discussed. It adopts the wideband dielectric unit-cell already introduced in [12] are investigated which consists of three perforated layers: the central presents a square hole, which is used to control the phase of the transmission coefficient, while the external layers are equal and have a truncated pyramid hole, which allow for improving the bandwidth and the efficiency. This cell element was firstly introduced in [10–13], which consists of three perforated layers: the central presents a square hole, which is used to control the phase of the transmission coefficient, while the external layers are equal and have a truncated pyramid hole, which allow for improving the bandwidth and the efficiency. This cell element was firstly introduced in [10], where its wideband behavior was tested only with numerical simulations of a medium size TA, designed using a conventional dielectric material and without considering the possible limitations introduced by the manufacturing process. In [11], the enhanced performances of the same unit-cell were compared with those of a conventional unit-cell having just a single layer with a square hole; in this case, a material usable for the realization of the TA with an Additive Manufacturing (AM) technique was considered, since this seemed to be the only adoptable method for the TA fabrication, but also in that work the designed antenna was only numerically analyzed, neglecting the constraints imposed by the AM process. The results in [11] proved that the TA with the tapered unit-cells provided a 1-dB bandwidth doubled (27.6%) with respect to that of the TA using the single layer cells (13.4%). The unit-cell was instead experimentally validated in [12]: firstly, several numerical tests were conducted to find the optimum geometric parameters and transmitarray configuration at Ka-band; then, the unit-cell was modified to make it 3D-printable in view of the limitations introduced by AM techniques; finally, a $15.6\lambda_0 \times 15.6\lambda_0$ antenna was designed, manufactured with 3D-printing techniques and experimentally characterized in the anechoic chamber. All of the configurations considered in [10–12], were designed neglecting the effect of the direction of arrival of the incident field on the unit-cell and they are only center-fed transmitarrays that generate a main beam in the broadside direction. In [13], some very preliminary considerations on the possibility to realize a mechanically steerable transmitarray are summarized: two different unit-cells are considered, one of which is the dielectric one; small size TAs working at 5.6 GHz were designed without taking into account any manufacturing limitations and considering the possibility to design a bifocal configuration or to use a global optimizer, and then characterized numerically. In this paper, a systematic analysis of the scanning beam capabilities of the unit-cell is first performed and the obtained results are presented in Section 2; the unit-cell has then been used for the design of a medium size TA, considering the steering of the incident field varying between $-30°$ and $+30°$: the adopted design procedure is described in detail in Section 3. A prototype has been therefore manufactured and measured: the results collected in Section 4 prove that the TA provides excellent radiation performances for different scanning angles, still maintaining for each of them a large bandwidth behavior; the 1-dB bandwidth remains almost the same for the different pointing angles, which are very stable over the entire bandwidth. Finally, in Section 4, some concluding remarks and possible future actions devoted to further improve the transmitarray features are discussed.

2. Unit-Cell Analysis

The basic structure of the unit-cell (UC) used for the design of the beam scanning TA was described in detail in [12], where it was proposed as an alternative solution to improve the bandwidth in transmitarray antennas; however, for the sake of clearness, its main features are summarized in the following.

As illustrated in Figure 1, the UC consists of three overlapping layers, made of the same dielectric material. The central layer has a square hole, whose size d is varied to control the phase of the transmission coefficient (S_{21}). The two external elements are characterized by a truncated pyramid hole that connects the hole in the central layer to the external aperture on the unit-cell external top (bottom) side, whose size W is the same for all of the unit-cells. The three layer configuration can be described with the equivalent transmission line model shown in Figure 2. The comparison with the lateral view of the unit-cell in Figure 1b highlights that the mid layer behaves as a uniform transmission line with characteristic impedance Z_{sqh}, while the external identical layers are equivalent to tapered transmission lines whose characteristic impedance $Z^*_{tap}(z)$ varies linearly along the z-axis, and therefore they act as wideband impedance inverters that match Z_{sqh} to the free-space impedance (Z_0). Both Z_{sqh} and Z_{sqh} are related to the effective permittivity of the corresponding layer that in its turn depends on the relative dielectric constant of the material and the size of the hole. The dimension of the tapered section changes linearly with z, and, as a consequence, the characteristic impedance $Z^*_{tap}(z)$ follows the same linear law. As discussed in [10], a good matching can be realized choosing the lengths l_1 and l_3 of the tapered lines at least equal to $\lambda_g/2$, where λ_g is the effective wavelength in the tapered section.

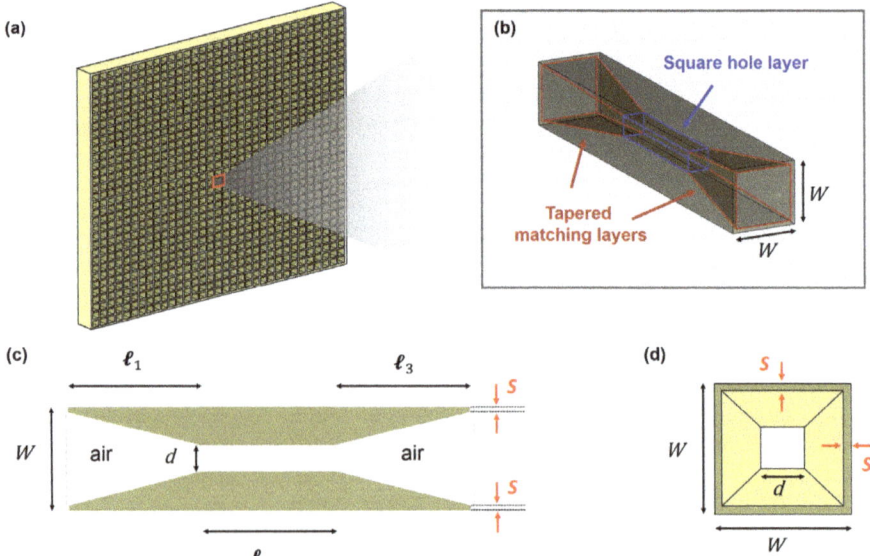

Figure 1. Sketch of a dielectric transmitarray and blow out of the unit-cell structure. (**a**) dielectric TA; (**b**) UC perspective view; (**c**) UC side view; (**d**) UC top view.

The unit-cell was designed to work in Ka-band, at a frequency $f_0 = 30$ GHz. To make possible its manufacturing with a 3D printer, it was designed using the commercial dielectric VeroWhite Plus (RGD835 provided by Stratasys®). This material was electromagnetically characterized using the waveguide method through which it was estimated that the average value of the dielectric constant in the considered frequency range is $\varepsilon_r = 2.77$ and the loss tangent is $\tan\delta = 0.021$. Considering the requirements introduced by the manufacturing process discussed in [12], and the parametric analysis performed to determine the optimal sizes of the unit-cell, its side is fixed to $W = 0.3\lambda_0 = 3$ mm, while the heights of each layer are $l_1 = l_2 = l_3 = 11$ mm, corresponding to a total cell thickness $T = l_1 + l_2 + l_3 = 3.3\lambda_0 = 33$ mm. Moreover, the resolution of the adopted 3D printer forces the maximum and minimum size of the holes to be equal to 2.65 and 0.5 mm, respectively; this choice does not allow for obtaining a perfect matching between the mid layer equivalent characteristic impedance and Z_0, with a consequent decrease of the amplitude of S_{21}, while a range of 360° is still achieved for its phase.

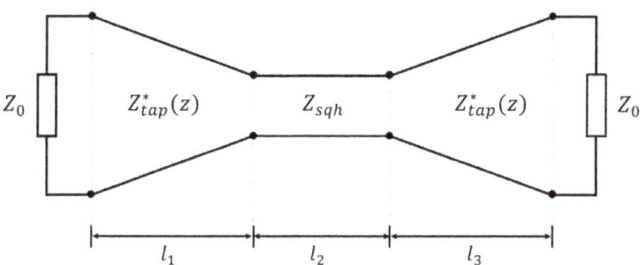

Figure 2. Equivalent transmission line model of the perforated dielectric unit-cell.

As also pointed out in [16], the scanning capabilities of a transmitarray depend, in the first instance, on the feature of the unit-cell, whose behavior must be as less as possible subject to the direction of arrival of the unit-cell, i.e., both the phase and the amplitude of S_{21} have to stay almost the same when the angle of incidence varies. In fact, this feature guarantees that, changing the angle of incidence, no higher modes resonate, as occurs for other unit-cell (see for instance [13]), degrading the antenna radiation performances and also that, when the direction of arrival changes, due to a variation of the position of the feed, the behavior of the unit-cell is not completely different from the presumed one. To study what happens in the unit-cell considered here, it has been analyzed using the Floquet excitations in CST MW Studio, used to compute the transmission coefficient as a function of d, for several angles of incidence θ_i (see inset in Figure 3a for its definition). The curves representing the variation of the amplitude and the phase of S_{21} are shown in Figure 3a,b, respectively. They have been calculated at f_0 and for $\theta_i = 10°, 20°, 30°, 40°$. Since the analyses for TE ($\phi_i = 0°$) and TM ($\phi_i = 90°$) modes provided the same results, only a single curve is plotted for each incidence angle. As it can be observed in Figure 3a, the amplitude of S_{21} is slightly affected by direction of arrival of the incident wave, with a reduction not greater than 0.5 dB when $\theta_i = 40°$. For what concerns the phase, from Figure 3b, it emerges that both its linearity and full range of variation are not affected by θ_i: in fact, the different curves are almost a translated copy of the other, with small changes in the slope. This means that the dielectric unit-cell behavior is not particularly sensitive to the angle of incidence, and this makes it a potential good candidate for the realization of a beam scanning TA, more suitable than other unit-cells, as some of those using metallic printed elements, where the oblique incident wave excites superior modes, at their turns responsible for a strong reduction in the amplitude of the transmission coefficient and discontinuities in the phase variation.

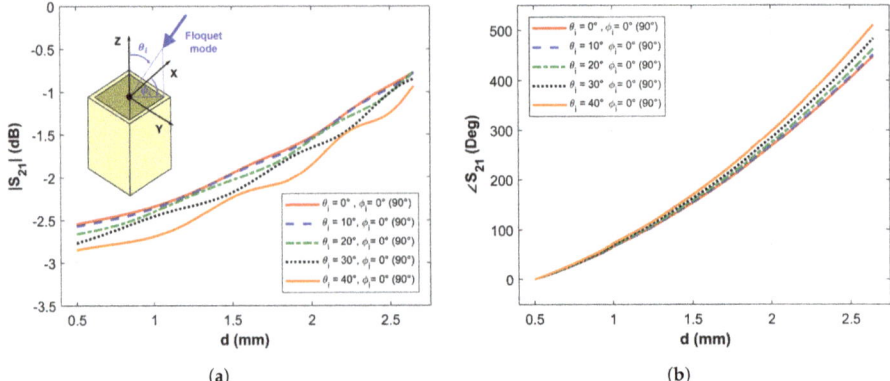

Figure 3. Comparison of the unit-cell transmission coefficient S_{21} computed at f_0 as a function of d for several angles of incidence θ_i: (a) amplitude; (b) phase.

3. Beam Scanning Transmitarray Design

Adopting the unit-cell described in the previous section, a medium-size transmitarray has been designed: it is composed by a matrix of 50×50 elements, corresponding to a size $D = 150$ mm $= 15\lambda_0$ at f_0. The first step in the design procedure concerns the evaluation of the required phase distribution. The transmission phase of each element is designed to compensate the spatial phase delay from the feed source to it [2]. Considering the coordinate system illustrated in Figure 4, the required phase distribution to obtain a focused beam in a direction normal to the surface is given by:

$$\psi_{req}(m,n) = k_0(\vec{r} - \vec{r}_{mn} \cdot \hat{u}_o) + \psi_o, \tag{1}$$

where k_0 is the propagation constant in free-space, $\vec{r}$ represents the distance from the feed to the (m,n)-th element, whose position is located by $\vec{r}_{mn}$, and $\hat{u}_o$ is the unit vector describing the main beam direction. The vector $\vec{r}_f$ determines the feed position, which is considered to be aligned to the center of the TA surface. The quantity ψ_o is a constant phase value, indicating that a relative phase rather than the absolute transmission phase is required for the transmitarray design. Using the array theory, the far-field radiation pattern for a rectangular aperture transmitarray of $M \times N$ elements can be predicted using the following approximated expression:

$$E(\theta,\phi) = \sum_{m=1}^{M} \sum_{n=1}^{N} \cos^{q_e}(\theta) \frac{\cos^{q_f}(\theta_f(m,n))}{|\vec{r}_{mn} - \vec{r}_f|} \cdot e^{-jk(|\vec{r}_{mn}-\vec{r}_f|-\vec{r}_{mn}\cdot\hat{u})} \cdot |S_{21}(m,n)|e^{j\angle S_{21}(m,n)}, \tag{2}$$

where q_e is the element power factor, q_f is the feed pattern power factor, $\theta_f(m,n)$ is the spherical angle in the feed coordinate system in correspondence of each element of the TA and for which the field radiated by the feed is computed. $|S_{21}(m,n)|$ and $\angle S_{21}(m,n)$ are the magnitude and phase of the transmission coefficient of each element, respectively. Once the required transmission phase $\psi_{req}(m,n)$ is determined for each element of the TA aperture, the corresponding element dimension is obtained using the curve that relates it to the phase $\angle S_{21}(m,n)$. This curve is generally obtained from the full-wave analysis of the unit-cell, as performed in Section 2. Typically, the evaluation $\angle S_{21}(m,n)$ is made under the assumption that all cells are illuminated by a field with a normal incidence. Actually, most of the elements are illuminated by oblique incidence angles, and therefore this approximation produces a phase error that depends on the degradation of the unit-cell performances with the angle θ_f. The incidence angle on each element of the aperture is given by

$$\theta_f(m,n) = acos\left(\frac{|\mathbf{r}|}{|\mathbf{r}_f|}\right). \tag{3}$$

The value of the incidence angles $\theta_f(m,n)$ in correspondence with the TA elements in the case considered here is shown in Figure 5a. It has been computed considering a focal length $F = 150$ mm ($F/D = 1$). The angle is maximum at the extremes of the aperture, achieving a peak value of about $34°$. The phase error introduced when the normal incidence is assumed for all the unit cells instead of the real direction of arrival of the incidence field is given by

$$\psi_{error}(m,n) = \angle S_{21}(m,n) - \angle S_{21}(\theta_f(m,n)), \tag{4}$$

where $\angle S_{21}(m,n)$ is the transmission phase computed with a normal incidence ($\theta_f = 0°$), while $\angle S_{21}(\theta_f(m,n))$ takes into account the different behavior of the UC with varying θ_f and it has been evaluated from the curves obtained in Figure 3b. In Figure 5b, the resulting phase error distribution $\psi_{error}(m,n)$ for the 50×50 TA is plotted: as it can be observed, at the center, the phase error is lower than $5°$ and it can be considered negligible, while beyond the phase jump, the error increases arriving up to about $30°$. It is higher when the position of the feed changes, as occurs during the mechanical beam scanning considered here. For instance, if we consider a scanning angle of $-30°$, the distribution of the incident angles becomes that in Figure 5c, in which the maximum angle achieved on the right corners is $51°$. In Figure 5d, it can be observed that the phase error related to this case has increased, reaching a maximum of about $45°$. Moreover, another important aspect has to be taken into account when the feed is rotating. If the Transmitarray is designed as a center-fed configuration, it is necessary to consider an additional phase error due to the difference between the required phase distribution needed when considering a different feed angle position and the required phase distribution for the center-fed case. Thus, the total phase error for scanning angles is the following:

$$\psi_{error}(m,n)|_{scan} = [\psi_{req}(m,n)|_{scan} - \psi_{req}(m,n)] + \psi_{error}(m,n), \tag{5}$$

where $\psi_{req}(m,n)$ is the required phase distribution for the center-fed case expressed by Equation (1), $\psi_{req}(m,n)_{scan}$ is the required phase shift needed for scanning angle case and $\psi_{error}(m,n)$ is the phase error related to the incidence angles by Equation (4). The second term of Equation (5) can be removed by choosing properly the values of the transmission phase $\angle S_{21}(\theta_f(m,n))$. This method has been applied to find the optimal phase distribution that allows for reducing the phase error for all the considered scanning angles.

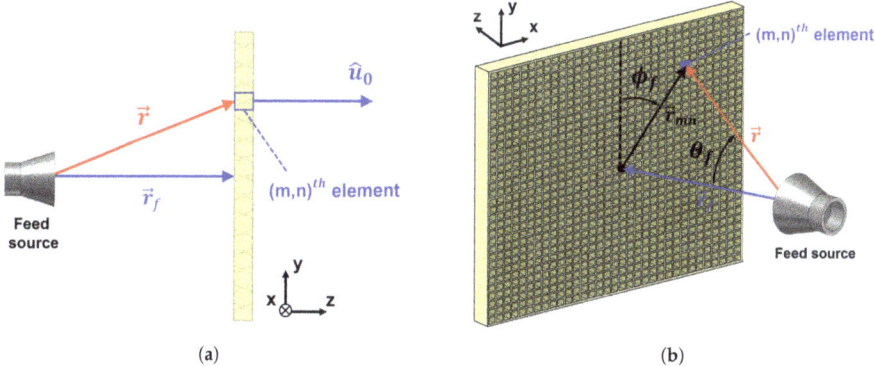

Figure 4. Sketch of the TA configuration with the adopted coordinate system: (a) side view; (b) perspective view of the illuminated surface.

The feed is a circular horn working in Ka-band: it has a gain of 17 dB at 30 GHz and its pattern can be modelled as $\cos^{q_f}(\theta)$ with $q_f = 12.5$. The focal distance between the horn and the dielectric structure is $F = 150$ mm ($F/D = 1$). Since here the main interest is in investigating the scanning capabilities of the TA, the beam steering that is obtained with a relative rotation between the two parts of the antenna, i.e., the transmitting surface and the feed, and this is equivalent to moving the horn along a circular arc having the center coincident with that of the TA itself, has been sketched in Figure 6a. In this way, the offset angle θ_f of the field impinging the TA changes, and consequently the direction of the main beam, is denoted with θ_b. A range of variation for θ_f of 60° in the E-plane has been considered for the transmitarray proposed here.

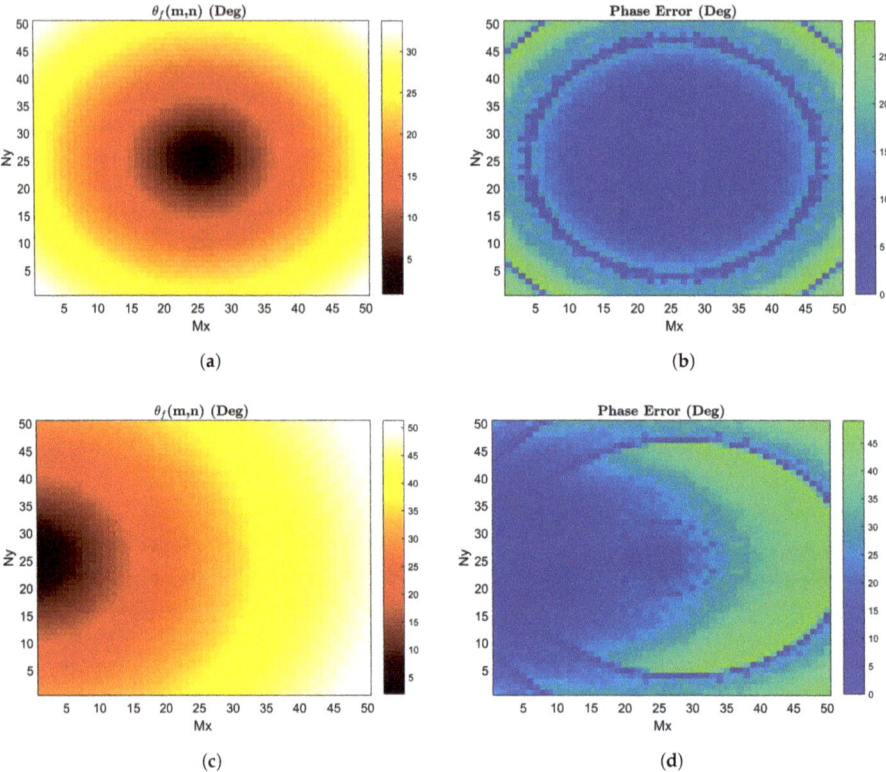

Figure 5. Matrix of the incidences angles $\theta_f(m,n)$ and related phase error: (a) incident angles for a center-fed TA; (b) phase error distribution considering a normal incidence; (c) incident angles when the scanning angle is $-30°$; (d) phase error distribution considering a normal incidence.

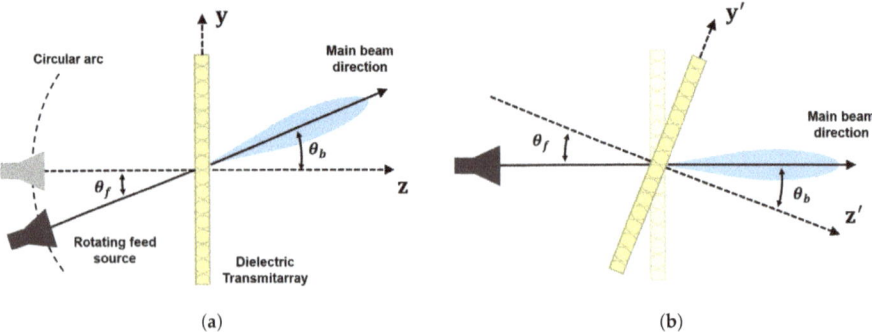

Figure 6. Pictorial view of the beam scanning mechanism (a) and of its actual implementation (b).

4. Prototype Manufacturing and Experimental Characterization

The performances of the TA have first been numerically analyzed using CST MW Studio; then, a prototype has been manufactured and experimentally characterized in an anechoic chamber. The prototype has been fabricated using the 3D-printer Object30 by Stratasys® that exploits PolyJet technology and allows for achieving a nominal resolution of 100 µm. In addition, the support connecting the feed to the TA and allowing its movement in the E-plane has been manufactured through a 3D-printing technique and, in particular, an FDM (Fused Deposition Modelling)-based 3D-printer has been exploited to realize it. A photo of the entire antenna and its measurement setup in the anechoic chamber is shown in Figure 7.

(a)   (b)

Figure 7. Beam scanning TA mounted in the anechoic chamber: (**a**) manufactured prototype and its support structure; (**b**) measurement setup including the open-ended waveguide probe.

Note that, to simplify the antenna assembly, the beam steering mechanism illustrated in Figure 6a is implemented in a slightly different way, as appears from Figure 7a more clearly from the sketch in Figure 6b: it is the TA that rotates around an axis in the horizontal plane, namely the x-axis in the reference coordinate system adopted in Figure 6, passing through the junctions that connect it to the supports; if this rotation is such that the angle between the y- and the y'-axis in the sketches of Figure 6b is equal to θ_f, the field radiated by the field impinges on the TA surface with an angle that is actually θ_f; in other words, it is sufficient to make a rotation of the coordinate system to overlap the drawing in Figure 6a to that in Figure 6b and this proves that they are equivalent. In the anechoic chamber, the measurement of the radiation pattern is done performing a 3D scanning in (θ, ϕ), where θ and ϕ are the angles used in the spherical coordinate system; when the measurement in the E-plane (i.e., the yz-plane in Figure 6) is performed, the radiation pattern as a function of θ is obtained. Therefore, θ_b is immediately found in the correspondence of the direction of maximum radiation and operating a translation proportional to the rotation applied to the Transmitarray, i.e., the angle θ_f.

The measured value of the main parameters characterizing the antenna radiation performances for the different θ_f are summarized in Table 1. In the first column, the value of these quantities for broadside radiation, where the gain is maximum, are listed. In the considered range for θ_f, the gain decreases less than 3 dB, from a maximum of 29.97 dB at 0° to 27.07 dB for $\theta_f = \pm 30°$. This corresponds to an enlargement of the main beam, that is, however, never larger than 1° in both planes (see 4th and 5th rows of the table). The decrease of the gain also affects the aperture efficiency, which reduces to 18% at the extremes of the scanning coverage. For what concerns the Side Lobe Levels (SLLs), their increase is more noticeable in the E-plane, where the beam is squinted, as also appears from the measured and simulated co-polarized radiation patterns at the frequency f_0 plotted in Figure 8, for several values of the angle θ_f. These patterns confirm the good scanning capabilities of the designed antenna, since they stay almost unchanged for scanning angles between -20° and 20°, but also, when $\theta_f = 30°$, they do not degrade significantly. Note that the maximum of the different plotted patterns occurs for angles θ_b that are slightly different from the corresponding angles θ_f. In fact, even if theoretically $\theta_b = \theta_f$, in practice, this condition is not verified especially when these angles increase and what happens is that θ_f is typically less than θ_b [28]. The ratio between the main scanning beam direction and the feed offset angle can be defined as the Beam Deviation Factor (BDF), i.e., BDF = θ_b/θ_f. For a very large scanning angle range, this parameter has to be optimized to obtain the desired performances. In the case of the designed TA, the values of the BDF are reported in the 7th row of Table 1: since it is equal to 0.9, it means that the achieved range of variation for θ_b is of 54°.

Table 1. Transmitarray performances for different scanning angles.

θ_f	0°	±10°	±20°	±30°
Gain	29.97 dBi	29.30 dBi	27.96 dBi	27.07 dBi
Ap. Eff. (30 GHz)	35.1%	30.1%	22.2%	18.1%
HPBW$_E$ (30 GHz)	3.8°	3.9°	4.1°	4.3°
HPBW$_H$ (30 GHz)	3.9°	3.9°	4.1°	4.8°
SLL$_E$ (30 GHz)	−22.6 dB	−20.4 dB	−16.0 dB	−11.1 dB
SLL$_H$ (30 GHz)	−22.6 dB	−22.4 dB	−20.1 dB	−15.5 dB
BDF	1	0.9	0.9	0.9
X-pol (30 GHz)	−40.4 dB	−39.6 dB	−38.5 dB	−36.3 dB
1-dB BW	23.5%	24.5%	23.3%	24%

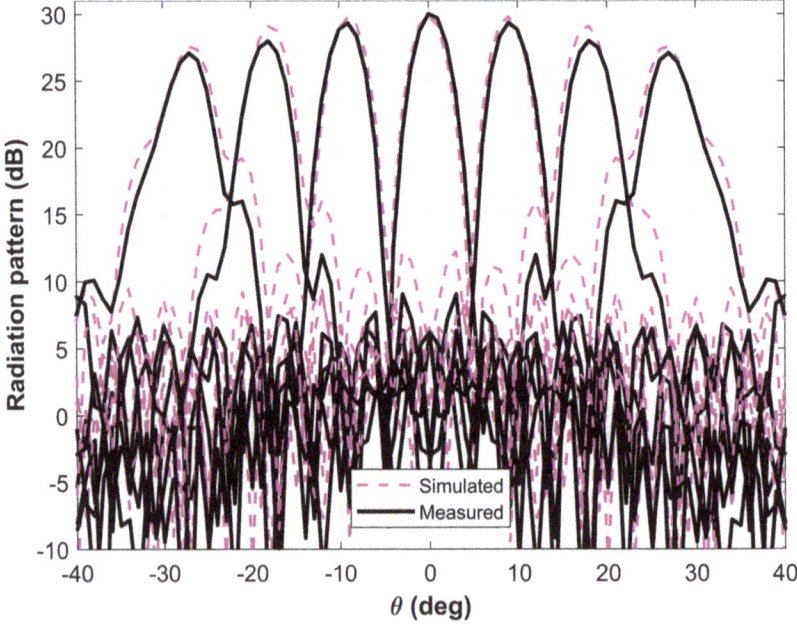

Figure 8. Simulated and measured co-polar radiation pattern in the E-plane at 30 GHz for different scanning angles.

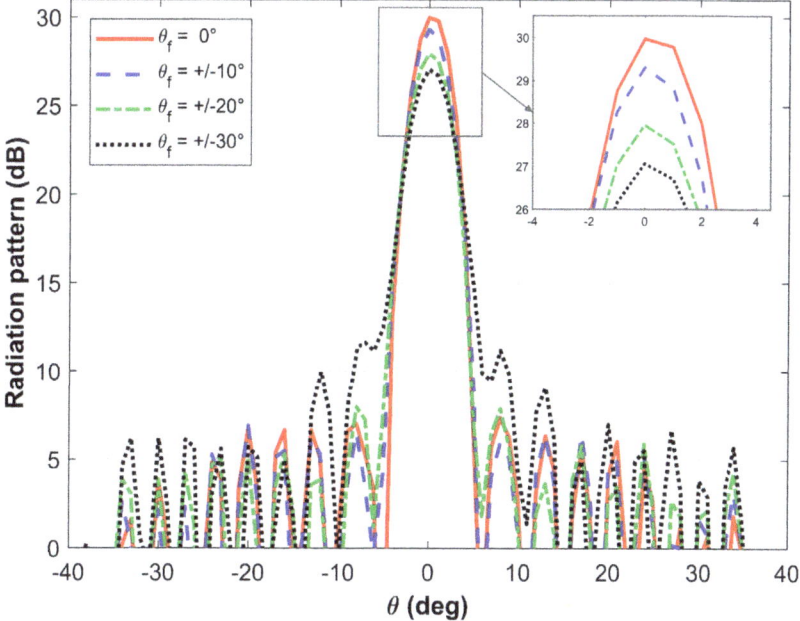

Figure 9. Measured co-polar radiation pattern in the H-plane at 30 GHz for different scanning angles.

Figure 9 shows the measured radiation patterns in H-plane, with a blow-up of the region close to the maximum in the inset. Apart from the reduction of the maximum gain, the radiation patterns are almost unchanged for $-20° \leq \theta_f \leq 20°$, and also for $\theta_f = \pm 30°$ the most significant effect is just a contained increase of the side lobes. In both Figures 8 and 9, only the co-polar component of the radiated field is shown, to not make the plot incomprehensible, but the values of the cross-pol measured at 30 GHz for each considered pointing angle are reported in Table 1: in all cases, it stays very low, reaching a maximum value of about -36 dB.

In Figure 10, the frequency variation of the gain for the different scanning angles is plotted, while the values of the 1-dB bandwidth are reported in the last row of Table 1. For the different values of θ_f, the 1-dB bandwidth remains larger than 23% and this represents an advantageous feature with respect to other transmitarray configurations implemented with metallic-dielectric elements, which have in most cases a narrow band and it decreases furthermore when the direction of maximum radiation changes. In Figure 11, the behavior of the measured and simulated gain at the design frequency as a function of the scanning angle is finally shown: the flatness of the curves is a further proof of the excellent scanning capabilities of the designed antenna; moreover, it is worth noticing the good agreement between the simulated and measured results.

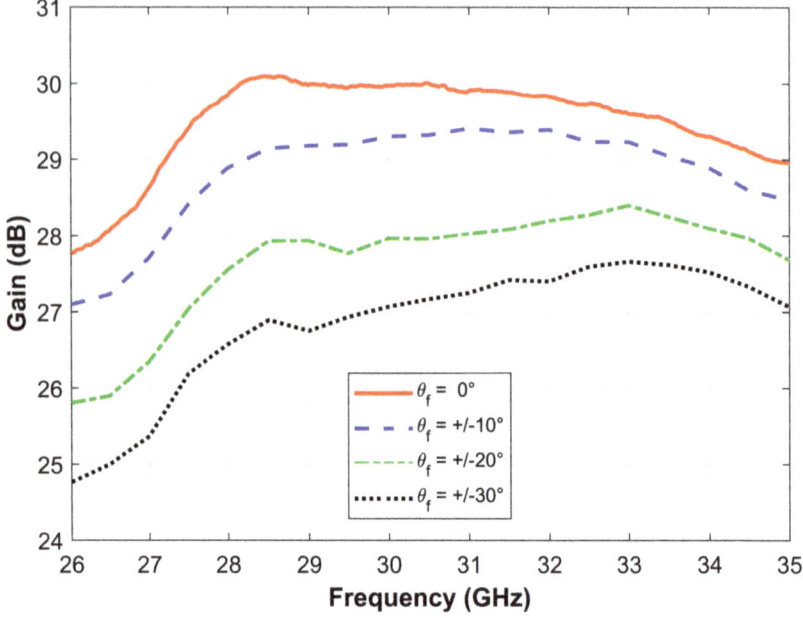

Figure 10. Measured frequency behaviour of the gain for different scanning angles.

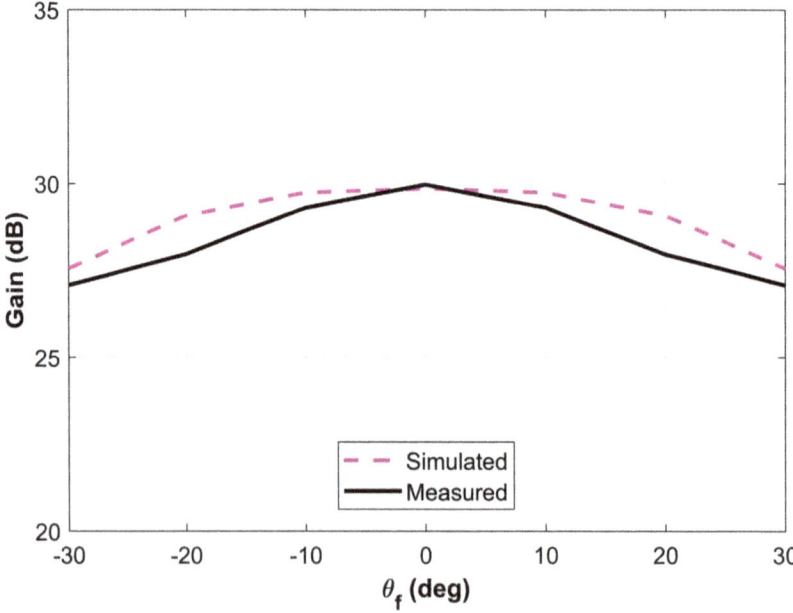

Figure 11. Comparison between the simulated and measured variation of the gain as a function of θ_f.

5. Conclusions

In this paper, the beam scanning capabilities of a 3D-printed dielectric transmitarray antenna have been analyzed and experimentally validated. The radiation patterns measured for different scanning angles in the range $\pm 27°$ showed a small variation of the gain, a controlled increase of the SLL and a wideband behavior, confirming the expected antenna properties. While these results prove the suitability of the considered unit-cell for the design of multibeam or beam scanning passive transmitarrays, some further optimization of the antenna would be necessary if similar performances would be obtained on larger scanning coverage. It will be the object of future investigations.

Author Contributions: Conceptualization, A.M. and P.P.; methodology, A.M. and P.P.; validation, G.D.; formal analysis, A.M.; investigation, A.M.; resources, P.P. and G.D.; data curation, A.M.; writing—original draft preparation, A.M. and P.P.; writing—review and editing, P.P.

Funding: This work was supported by the General Directorate for Cultural and Economic Promotion and Innovation of the Ministry of Foreign Affairs and International Cooperation, of the Italian Republic.

Acknowledgments: The authors thank the research group of Chilab for the antenna manufacturing.

Conflicts of Interest: The authors declare no conflict of interest.

References

1. Huang, J.; Encinar, J.A. *Reflectarray Antennas*; Wiley IEEE Press: Hoboken, NJ, USA, 2008.
2. Abdelrahman, A.H.; Yang, F.; Esherbeni, A.Z.; Nayeri, P. *Analysis and Design of Transmitarray Antennas*; M&C Publishers: San Francisco, CA, USA, 2017.
3. Ryan, C.G.M.; Chaharmir, M.R.; Shaker, J.R.B.J.; Bray, J.R.; Antar, Y.M.; Ittipiboon, A. A wideband transmitarray using dual-resonant double square rings. *IEEE Trans. Antennas Propag.* **2010**, *58*, 1486–1493. [CrossRef]
4. Abdelrahman, A.H.; Elsherbeni, A.Z.; Yang, F. High-gain and broadband transmitarray antenna using triple-layer spiral dipole elements. *IEEE Antennas Wirel. Propag. Lett.* **2014**, *13*, 1288–1291. [CrossRef]
5. Cai, Y.M.; Li, W.; Li, K.; Gao, S.; Yin, Y.; Zhao, L.; Hu, W. A Novel Ultra-Wideband Transmitarray Design Using Tightly Coupled Dipole Elements. *IEEE Trans. Antennas Propag.* **2019**, *67*, 242–250. [CrossRef]
6. Abdelrahman, A.H.; Esherbeni, A.Z.; Yang, F. Transmitarray antenna design using cross slot elements with no dielectric substrate. *IEEE Antennas Wirel. Propag. Lett.* **2014**, *13*, 177–180. [CrossRef]
7. Rahmati, B.; Hassani, H.R. High-efficient wideband slot transmitarray antenna. *IEEE Trans. Antennas Propag.* **2015**, *63*, 5149–5155. [CrossRef]
8. Mahmoud, A.-E.; Hong, W.; Zhang, Y.; Kishk, A. W-band mutlilayer perforated dielectric substrate lens. *IEEE Antennas Wirel. Propag. Lett.* **2014**, *13*, 734–737. [CrossRef]
9. Yi, H.; Qu, S.W.; Ng, K.B.; Chan, C.H.; Bai, X. 3-D printed millimeter-wave and terahertz lenses with fixed and frequency scanned beam. *IEEE Trans. Antennas Propag.* **2016**, *64*, 442–449. [CrossRef]
10. Massaccesi, A.; Pirinoli, P. Enhancing the bandwidth in transmitarray antennas using tapered transmission line matching approach. In Proceedings of the 12th European Conference on Antennas and Propagation (EuCAP 2018), London, UK, 9–13 April 2018.
11. Massaccesi, A.; Pirinoli, P.; Vardaxoglou, J.C. A multilayer unit-cell for perforated dielectric transmitarray antennas. In Proceedings of the 2018 IEEE International Symposium on Antennas and Prop & USNC/URSI National Radio Science Meeting, Boston, MA, USA, 8–13 July 2018; pp. 263–264.
12. Massaccesi, A.; Pirinoli, P.; Bertana, V.; Scordo, G.; Marasso, S.L.; Cocuzza, M.; Dassano, G. 3D-Printable dielectric transmitarray with enhanced bandwidth at millimiter-waves. *IEEE Access* **2018**, *6*, 46407–46418. [CrossRef]
13. Beccaria, M.; Massaccesi, A.; Pirinoli, P. Multibeam transmitarrays for 5G antenna systems. In Proceedings of the 2018 IEEE Seventh International Conference on Communications and Electronics (ICCE), Hue, Vietnam, 18–20 July 2018; pp. 217–221.
14. Jiang, M.; Chen, Z.N.; Zhang, Y.; Hong, W.; Xuan, X. Metamaterial-based thin planar lens antenna for spatial beamforming and multibeam massive MIMO. *IEEE Trans. Antennas Propag.* **2017**, *65*, 464–472. [CrossRef]

15. Hou, Y.; Chang, L.; Li, Y.; Zhang, Z.; Feng, Z. Linear Multibeam Transmitarray Based on the Sliding Aperture Technique. *IEEE Trans. Antennas Propag.* **2018**, *66*, 3948–3958. [CrossRef]
16. Liu, G.; Pham, K.; Cruz, E.M.; Ovejero, D.G.; Sauleau, R. A millimeter wave transparent transmitarray antenna using meshed double circle rings elements. In Proceedings of the 12th European Conference on Antennas and Propagation (EuCAP 2018), London, UK, 9–13 April 2018.
17. Lima, E.B.; Matos, S.A.; Costa, J.R.; Fernandes, C.A.; Fonseca, N.J.G. Circular polarization wide-angle beam steering at Ka-band by in-plane translation of a plate lens antenna. *IEEE Trans. Antennas Propag.* **2015**, *63*, 5443–5455. [CrossRef]
18. Liu, G.; Cruz, E.M.; Pham, K.; Ovejero, D.G.; Sauleau, R. Low Scan Loss Bifocal Ka-band Transparent Transmitarray Antenna. In Proceedings of the 2018 IEEE International Symposium on Antennas and Prop & USNC/URSI National Radio Science Meeting, Boston, MA, USA, 8–13 July 2018; pp. 1449–1450.
19. Clemente, A.; Dussopt, L.; Sauleau, R.; Potier, P.; Pouliguen, P. Wideband 400-Element Electronically Reconfigurable Transmitarray in X Band. *IEEE Trans. Antennas Propag.* **2013**, *61*, 5017–5027. [CrossRef]
20. Pan, W.; Huang, C.; Ma, X.; Jiang, B.; Luo, X. A dual linearly polarized transmitarray element with 1-Bit phase resolution in X-band. *IEEE Antennas Wirel. Propag. Lett.* **2015**, *14*, 167–171. [CrossRef]
21. Di Palma, L., Clemente, A., Dussopt, L., Sauleau, R., Potier, P., Pouliguen, P. Circularly-polarized reconfigurable transmitarray in Ka-band with beam scanning and polarization switching capabilities. *IEEE Trans. Antennas Propag.* **2017**, *65*, 529–540. [CrossRef]
22. Nguyen, D.B.; Pichot, C. Unit-Cell loaded with PIN diodes for 1-bit linearly polarized reconfigurable transmitarrays. *IEEE Antennas Wirel. Propag. Lett.* **2019**, *18*, 98–102. [CrossRef]
23. Padilla, P.; Munoz-Acevedo, A.; Sierra-Castaner, M.; Sierra-Perez, M. Electronically reconfigurable transmitarray at Ku band for microwave applications. *IEEE Trans. Antennas Propag.* **2010**, *58*, 2571–2579. [CrossRef]
24. Lau, J.Y.; Hum, S.V. A wideband reconfigurable transmitarray element. *IEEE Trans. Antennas Propag.* **2012**, *60*, 1303–1311. [CrossRef]
25. Lau, J.Y.; Hum, S.V. Reconfigurable transmitarray design approaches for beamforming applications. *IEEE Trans. Antennas Propag.* **2012**, *60*, 5679–5689. [CrossRef]
26. Sazegar, M.; Yuliang, Z.; Kohler, C.; Maune, H.; Nikfalazar, M.; Binder, J.R.; Jakoby, R. Beamsteering transmitarray using tunable frequency selective surface with integrated ferroelectric varactors. *IEEE Trans. Antennas Propag.* **2012**, *60*, 5690–5699. [CrossRef]
27. Huang, C.; Pan, W.; Ma, X.; Jiang, B.; Cui, J.; Luo, X. Using reconfigurable transmitarray to achieve beam-steering and polarization manipulation applications. *IEEE Trans. Antennas Propag.* **2015**, *63*, 4801–4810. [CrossRef]
28. Rengarajan, S.R. Scanning and defocusing characteristics of microstrip reflectarray. *IEEE Antennas Wirel. Propag. Lett.* **2010**, *9*, 163–166. [CrossRef]

© 2019 by the authors. Licensee MDPI, Basel, Switzerland. This article is an open access article distributed under the terms and conditions of the Creative Commons Attribution (CC BY) license (http://creativecommons.org/licenses/by/4.0/).

Article

Design of a SIW Variable Phase Shifter for Beam Steering Antenna Systems

Sara Salem Hesari * and Jens Bornemann

Department of Electrical and Computer Engineering, University of Victoria, Victoria, BC V8W 2Y2, Canada; jbornema@ece.uvic.ca
* Correspondence: ssalem@uvic.ca; Tel.: +1-250-721-8989 (ext. 2769)

Received: 26 July 2019; Accepted: 8 September 2019; Published: 11 September 2019

Abstract: This paper proposes a new beam steering antenna system consisting of two variable reflection-type phase shifters, a 3 dB coupler, and a 90° phase transition. The entire structure is designed and fabricated on a single layer of substrate integrated waveguide (SIW), which makes it a low loss and low-profile antenna system. Surface mount tuning varactor diodes are chosen as electrical phase control elements. By changing the biasing voltage of the varactor diodes in the phase shifter circuits, the far-field radiation pattern of the antenna steers from −25° to 25°. The system has a reflection coefficient better than −10 dB for a 2 GHz bandwidth centered at 17 GHz, a directive radiation pattern with a maximum of 10.7 dB gain at the mid-band frequency, and cross polarization better than 20 dB. A prototype is fabricated and measured for design verification. The measured far-field radiation patterns, co and cross polarization, and the reflection coefficient of the antenna system agree with simulated results.

Keywords: beam steering antenna system; SIW technology; variable phase shifter; varactor diodes

1. Introduction

Beam steering antenna systems have become progressively popular in modern communications. Transferring and receiving high rate data, scanning a large scale of the atmosphere, having compact structures with minimum losses—all are attributed to and achieved through electrical beam steering antenna systems.

The main components in designing a beam steering antenna system are the phase shifters. Phase shifting in an antenna array can be achieved by using different methods. Included are variable phase shifters equipped with varactor diodes [1], PIN diode phase shifters (which have three types: switched line, loaded line, and reflection) [2], a Nolen matrix [3] or a Butler matrix [4] for constant phase shifts, complementary split ring resonators (CSRRs) [5], a tunable band-pass filter as phase shifter [6], a liquid crystal miniaturized phase shifter [7], some innovative vector-sum phase shifters by using CMOS technology [8,9], and a continuously steered beam in a range of ±45° through an array of vector controllers [10], to name only a few.

Substrate integrated waveguide (SIW) has recently attracted a lot of attention because of its compactness, low loss, and its promising performance in mm-wave and microwave applications. Therefore, researchers started to not only use this technology for passive components, but also in active component areas like variable phase shifters. Reference [1] introduces two SIW phase shifters. First, an inline phase shifter using a varactor diode with a phase shift of 25°. Secondly, a 180° SIW reflection-type phase shifter including a biasing circuit, four varactor diodes, and two open stubs. By changing the loading of a waveguide, a controllable phase shifter is achieved in [11]. The phase shift is controlled by four PIN diodes inserted in the substrate, and a maximum phase shift of 45° is obtained. An EM-wave coupling technique is proposed for achieving a wider bandwidth in a two layer

SIW phase shifter design [12], and a phase shift of up to 180° is obtained by using eight PIN diodes. The combination of a 90° hybrid consisting of two waveguide layers and four varactor diodes, two on top and two on the bottom, results in a four layer structure. This phase shifter has nearly 360° phase variation [13]. The authors of [14] propose an SIW K-band phase shifter with minimum 105° phase tuning range and 4 dB insertion loss by using varactor diodes as tuning elements. Note that since varactor diodes and their connecting transitions add losses to the circuit and also affect the radiation pattern of the antenna due to radiations from the connecting microstrip lines, the use of fewer varactor diodes is desirable.

A 2.8 GHz to 4.8 GHz beam steering microstrip antenna for radar applications is introduced in [15]. Beam steering is achieved with a pair of PIN diodes connecting the stubs to the ground and changing the state between On and Off. A single narrow bandwidth patch antenna is proposed in [5] which steers the beam by using complementary split ring resonators (CSRRs) as part of the ground plane. The antenna with double CSRR loading is able to steer the beam from −51° to 48°. A scan range up to 16° is obtained with an integrated 95 mm diameter lens antenna [16] in E-band. The beam is directed to different angles by switching between the feed elements of the array. Reference [2] presents a reconfigurable beam steering system comprising four patch antennas, three power dividers, and four phase shifters. A switched line PIN diode phase shifter is used for directing the beam between −30° to 30°. Reference [17] introduces a new approach to generate beam steering antenna systems without using phase shifters. A 4 × 4 planar circularly polarized spiral antenna array is designed and verified in a spherical coordinate system. As ϕ changes from 0 to 2π, the phase of the CP antenna's co-pol radiation will change by 2π, while r and θ remain constant. A single layer SIW version of a 4 × 4 Butler matrix system is presented in [18], achieving scanning angles of ±25°. A broadband beam steering antenna for 5G applications is introduced in [4]. This design includes a 4 × 4 Butler matrix which applies progressive phase shift between microstrip antenna elements. Since a Butler matrix has some constant phase differences at the output, a signal can be directed only to some specific angles. Reference [6] demonstrates a new method of using band-pass filters as phase shifters or tunable filters by using coupled microstrip resonators loaded with a varactor diode in the filter structure. A 1 × 4 steerable dielectric resonator antenna array at C / X band is presented in [19]. Inkjet-printed barium strontium titanate (BST) thick-film is applied as phase shifters. Integrated metal-insulator-metal varactors are used for tuning the phase. This antenna system achieves a ±30° beam steering range.

Firstly, this paper proposes an SIW Ku-band variable phase shifter. This phase shifter is a combination of a 3 dB coupler with an isolation better than 30 dB at the mid-band frequency of 17 GHz and 22 dB in the entire operating bandwidth, with two varactor diodes as electrical phase control elements. Then, an antenna array consisting of two Vivaldi antennas is introduced for designing a beam steering antenna system. The proposed beam steering system is designed on a single SIW layer which makes it a low-profile structure with minimum losses. The system has a 2 GHz bandwidth with a reflection coefficient better than −10 dB between 16 GHz and 18 GHz and 50° controllable beam steering capability. Therefore, the design criteria consisting of reflection coefficient better than 10 dB, 10 dB gain, 50° scanning angle with cross polarization better than 20 dB, which are normally required for tracking systems and other possible scanning applications, are achieved successfully.

2. SIW Variable Phase Shifter

SIW phase shifters have the advantages of being compact, low loss, and planar which make them suitable for integration with other components on a single layer. Phase shifters are critical components in the mm-wave and microwave area for applications like phased array systems in radio astronomy, and beam steering systems for tracking applications in wireless and satellite communications. Phase shifters divide into two main groups: Fixed phase shifters and variable phase shifters. This paper's focus is on the design of a variable SIW phase shifter.

2.1. SIW 3-dB Coupler

A hybrid coupler with high isolation and 3 dB coupling is designed on RT/duroid 6002 substrate with relative permittivity of 2.94, thickness of $h = 0.508$ mm, and loss tangent of 0.0012 [20]. The component is designed and simulated in CST Microwave Studio 2018. Figure 1 presents the SIW 3 dB coupler structure. The required coupling and isolation are achieved by optimizing the length of the gap in the middle, the location of the vias on both sides, and also the middle vias. As shown in Figure 2a, this design has a reflection coefficient better than −22 dB in the entire frequency range. It has an isolation between the adjacent ports of higher than 30 dB at mid-band frequency and 3 dB power division between 16 GHz and 18 GHz. Figure 2b demonstrates that the transmission coefficient, when including dielectric and metallic losses in the analysis, amounts to an insertion loss of less than 0.5 dB.

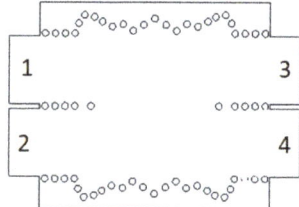

Figure 1. Substrate integrated waveguide 3 dB coupler.

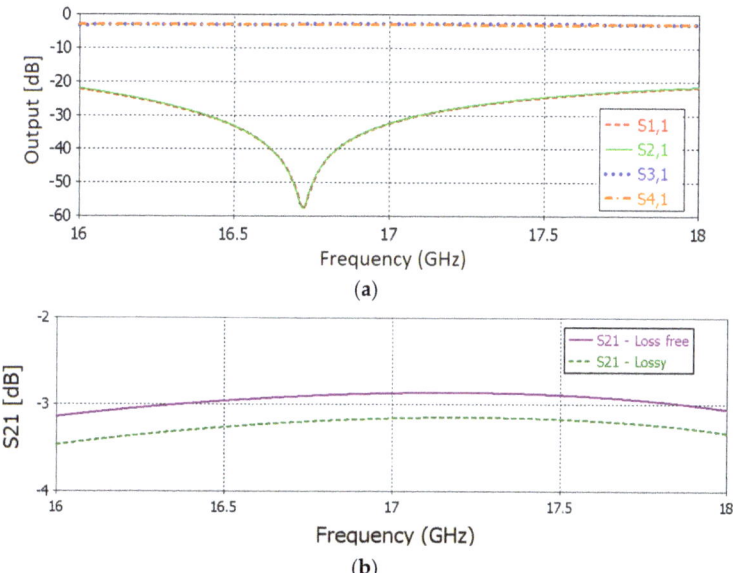

Figure 2. Scattering parameters of the 3 dB coupler, (a) ideal, (b) $S_{21} = S_{31}$ with dielectric and metal losses included compared to lossless analysis.

2.2. SIW Variable Phase Shifter

A reflection-type variable phase shifter is designed in this section, which consists of an SIW 3-dB coupler and the biasing circuit for the varactor diodes. Figure 3 shows the structure and parameters of the phase shifter. The phase shifter parameters including the coupler are designed based on the theoretical analysis introduced in [21]. The reader is referred to [21] for details. The waveguide width is designed based on the cutoff frequency [14], and the SIW width is defined by using the proposed formula in [22]. The *d/p* ratio is chosen as 0.65 to have minimum wave leakage in the SIW circuit. Table 1 presents the phase shifter's dimensions.

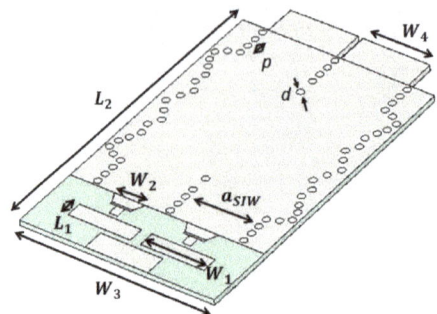

Figure 3. SIW variable phase shifter structure and parameters.

Table 1. Dimensions of the SIW variable phase shifter.

Parameter	Dimension (mm)	Parameter	Dimension (mm)
p	1	W_1	6
d	0.65	W_2	3
L_1	1.5	W_3	18.6
L_2	29.46	W_4	6.508
h	0.508	a_{SIW}	7

Before applying the varactor diodes to the coupler introduced in Section 2.1, two capacitors are applied to ports 3 and 4 (Figure 1) to analyze the phase variation that can be obtained by this circuit. As revealed in Figure 4, by changing the capacitor value from 0.1 pF to 5 pF, a phase variation of about 125° is obtained. After 5 pF, the value of the capacitor is so high that it resembles a short circuit (about 2 Ω impedance), thus the phase variation saturates.

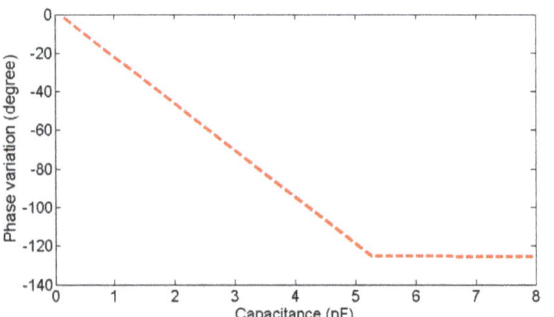

Figure 4. Phase variation of the 3-dB coupler by applying different capacitance values at ports 3 and 4.

For this design, the silicon hyper abrupt junction varactor diode SMV2201-040LF [23] is chosen, which is specifically designed for wideband applications. For examining the performance of this diode, its equivalent circuit is modeled and analyzed in CST, which is demonstrated in Figure 5. A microstrip-to-SIW transition comprising a 50-Ω microstrip line and a tapered microstrip section as well as two parasitic compensation stubs with width W_1 and length L_1 (Figure 3) are applied to ports 3 and 4 of the coupler for achieving the required capacitance for a 125° phase shift. An inductor is inserted between the DC voltage and the varactor diode to provide a quasi-open circuit for the RF signal. Table 2 presents the relationship between the output phase of the phase shifter circuit and biasing voltage of the varactor diodes for a single-phase shifter. By increasing the biasing voltage from 0 to 20 V, a nonlinear variation of phase difference between 0° and 126° is achieved.

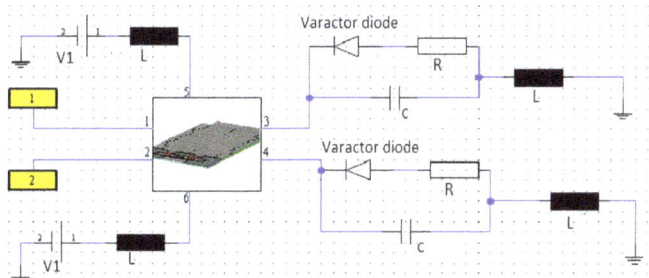

Figure 5. Schematic part of the SIW variable phase shifter and equivalent circuits of the varactor diodes.

Table 2. Phase shifter angle and capacitance as a function of biasing voltage.

Voltage (V)	Capacitor (pF)	Phase Shifter Angle (degree)
0	2.10	0
5	0.67	45.51
10	0.33	75.92
15	0.26	106.34
20	0.23	126.02

Figures 6–8 demonstrate the performance of the variable phase shifter by applying 0 to 20 V biasing voltage to the circuit. While applying five different voltages, Figure 6 shows the reflection coefficient of the variable phase shifter which is better than −15 dB in all cases between 16 GHz and 18 GHz. The phase variation achieved by the equivalent circuit of the varactor diode is demonstrated in Figure 7 and verifies the 120° phase differences. Figure 8 presents the insertion loss of the proposed phase shifter, which improves from 11 dB to 3 dB by increasing the biasing voltage from 0 to 20 V.

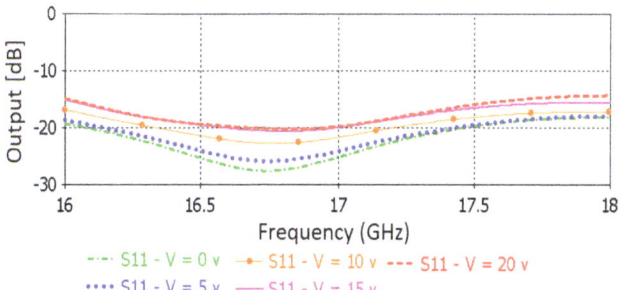

Figure 6. Reflection coefficient of the SIW variable phase shifter for five different biasing voltages.

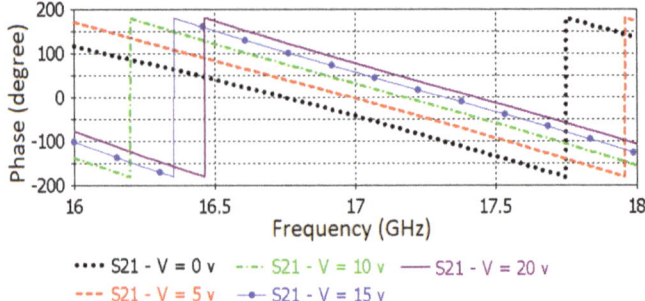

Figure 7. Phase variation of the SIW variable phase shifter for five different biasing voltages.

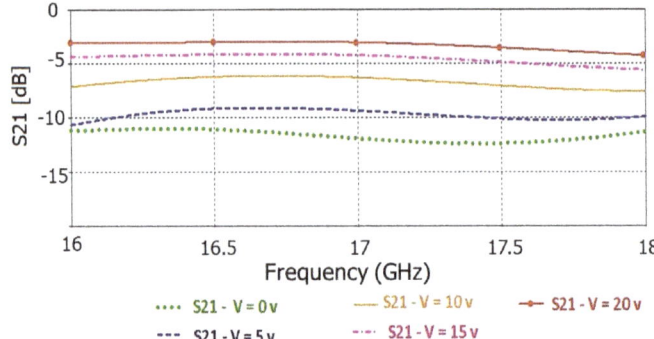

Figure 8. Insertion loss of the SIW variable phase shifter for five different biasing voltages.

3. SIW Beam Steering Antenna System

By using electrically variable phase shifters, we are able to change the phase of the signal in any individual antenna element in an array, which consequently steers the far-field radiation pattern of the array in space without the need for any mechanical rotation.

3.1. Vivaldi Antenna

An antipodal Vivaldi antenna is employed as the radiating element for this antenna system. It is designed based on the parametric study in [24], and then fine optimized [25] for performance in the required frequency band of 16 GHz to 18 GHz. The Vivaldi antenna is horizontally polarized, and vertical polarization is blocked by the comb-like corrugations on two arms of the antenna. Therefore, the corrugations improve cross polarization. The proposed antenna has a wide bandwidth with a reflection coefficient better than −15 dB in the entire operating bandwidth, as presented in Figure 9a. Its end-fire radiation pattern with high directivity is displayed in Figure 9b. This single antenna provides about 9.6 dB gain at mid-band frequency. The proposed Vivaldi antenna has a very directive radiation pattern with 50° HPBW, which consequently limits the scan range of the beam steering system to about 50°.

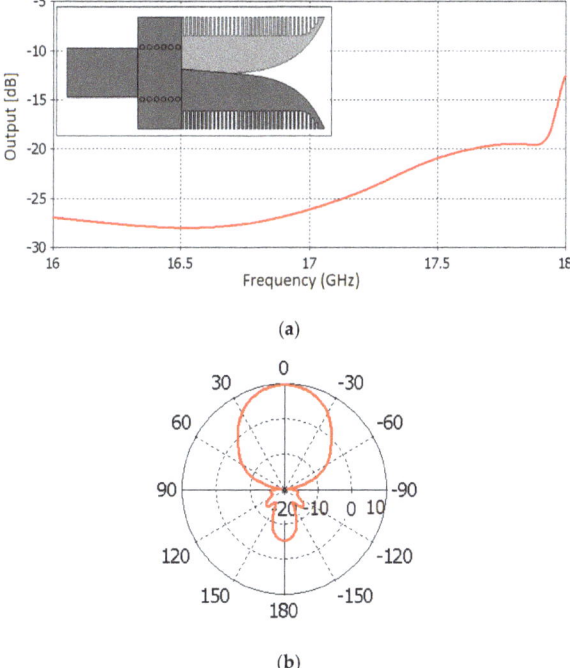

Figure 9. SIW Vivaldi antenna, (**a**) reflection coefficient, (**b**) far-field radiation pattern at 17 GHz.

3.2. Beam Steering Antenna System

Beam steering is accomplished by applying a progressive phase shift between neighboring elements in an array. The proposed beam steering antenna system includes two variable phase shifters, an antenna array consisting of two Vivaldi antennas, and a 90° phase transition. A microstrip-to-SIW transition is designed for feeding the antenna system by K-connectors at ports 1 and 2. First, a 50-Ω microstrip line is designed based on principles explained in [26]. Then, it is tapered out for a better match. Figure 10 demonstrates the entire beam steering antenna system.

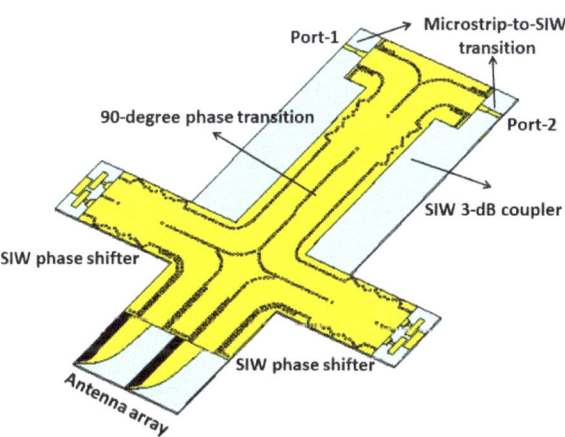

Figure 10. Beam steering antenna system.

A SIW 90° phase transition is applied between the 3-dB coupler and phase shifters. This transition causes 0° and 180° phase differences between the input ports of the right and left phase shifter when exciting port 1 and port 2, respectively [27]. Therefore, a 0° (sum) and 180° (difference) steering range is obtained by using this transition. Figure 11 displays the electric field distribution in the entire antenna system when port 1 is excited. As exhibited, minimum loss and no wave-leakage through the SIW transitions are accomplished.

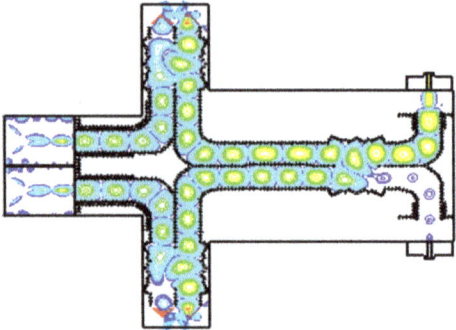

Figure 11. Electric field distribution in SIW beam steering antenna system at 17 GHz when port 1 is excited.

In order to calculate the beam steering range of the system by considering a phase shifter with a maximum of 125° phase variation, the following formula is used [28]:

$$\psi = \frac{-360f}{c}\left([d_x + \Delta_x]\sin\theta\cos\phi + d_y\sin\theta\sin\phi\right) \quad (1)$$

where ψ is the phase shift between the elements, f is the frequency, c is the speed of light, d_x and d_y are x- and y-element spacing, Δ_x is the row offset for a non-rectangular lattice, and θ and ϕ are the beam steering angles.

As revealed in [23], by applying 0 to 20 V bias to the varactor diode, its capacitance changes between 0.23 pF and 2.1 pF. The gain pattern of the beam steering antenna system is presented in Figure 12 at mid-band frequency of 17 GHz, while port 1 is excited and port 2 is terminated. As expected by applying the same biasing voltage to both right and left phase shifters, a signal with a 180° phase difference feeds the Vivaldi antennas which consequently causes a deep null in the broad-sight direction of the antenna.

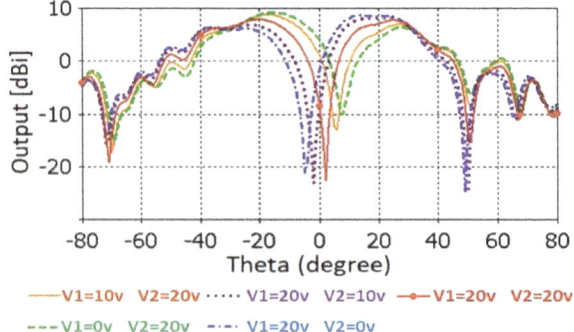

Figure 12. Simulated gain patterns of the beam steering antenna system at 17 GHz for 5 different loadings while port 1 is excited.

Figure 13 presents the gain pattern of the antenna system for five different loadings when port 2 is excited and port 1 is terminated. Due to the 90° transition by exciting port 2, signals go through the SIW phase shifters with 180° phase differences. Therefore, by applying the same voltage to the right and left phase shifters, another 180° phase difference is added to the signal which causes a directive radiation pattern in the broad-sight direction of the system.

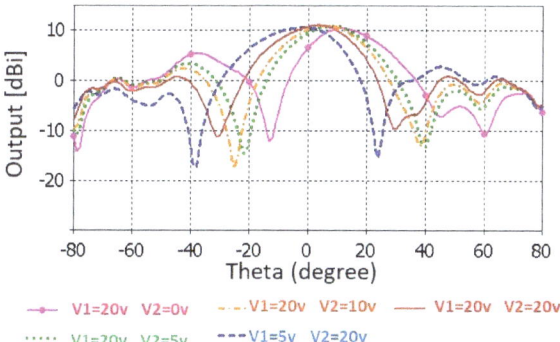

Figure 13. Simulated gain pattern of the beam steering antenna system at 17 GHz for 5 different loadings while port 2 is excited.

By considering the progressive phase shift between the individual antennas by exciting port 1 and port 2, a beam steering range from −25° to +25° is obtained, which verifies the calculated value by using Equation (1).

4. Experimental Results

The proposed SIW antenna system is fabricated on a single layer of RT/duroid 6002. The top and bottom views of the fabricated prototype and details of the phase shifter circuit are displayed in Figure 14. Two end launch K-connectors are used for exciting the ports. Four varactor diodes are soldered to the board between the microstrip and compensation stubs. Two inductors are placed in series with the varactor diodes and DC voltage supply which can be replaced by chip inductors.

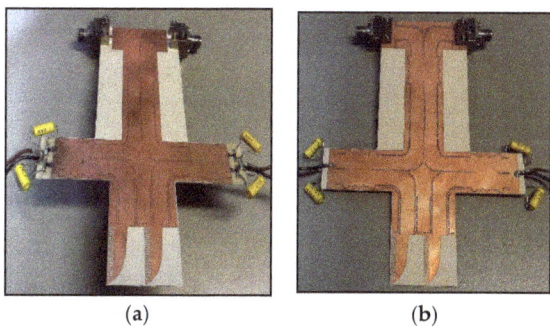

(a)   (b)

Figure 14. Cont.

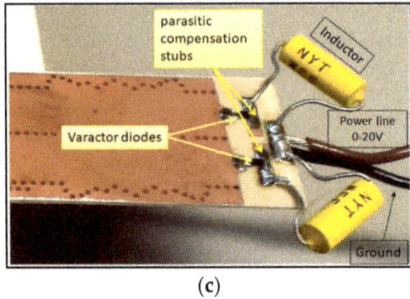

(c)

Figure 14. Beam steering antenna system prototype, (**a**) top view, (**b**) bottom view, (**c**) details of the phase shifter circuit.

For performance verification, the antenna system is measured in a far-field anechoic chamber using an Anritsu 37397C vector network analyzer. Figure 15 presents both simulated and measured reflection coefficients of the beam steering antenna system for ports 1 and 2. The reflection coefficient is better than −10 dB in the entire frequency range, and measurements show a fairly good agreement with simulation results.

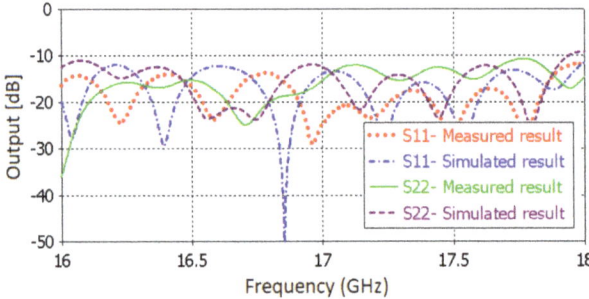

Figure 15. Measured and simulated reflection coefficients of the beam steering antenna system.

The far-field radiation pattern of this antenna system is measured for three different loadings for performance confirmation. V_1 and V_2 are the biasing voltages applied to the left and right phase shifters, respectively. First, $V_1 = 0$ V and $V_2 = 20$ V are applied, which steers the null to −8°, then $V_1 = 20$ V and $V_2 = 10$ V are applied, which steers the null to 5°, and finally $V_1 = V_2 = 20$ V is applied, which has the null at 0° and the main lobes at ± 25° which are shown in Figure 16.

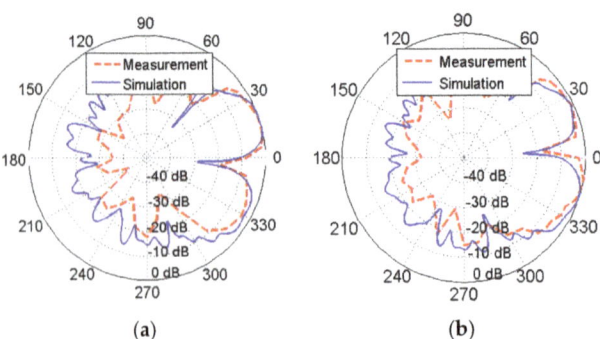

Figure 16. *Cont.*

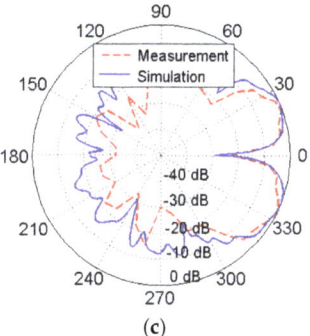

Figure 16. Measured and simulated far-field radiation pattern of the antenna system while exciting port 1 with (**a**) $V_1 = 0$ V and $V_2 = 20$ V, (**b**) $V_1 = 20$ V and $V_2 = 10$ V, (**c**) $V_1 = V_2 = 20$ V.

The co and cross polarization of this antenna system are measured by exciting port 2, terminating port 1, and applying the 20 V biasing voltage to both phase shifters. Figure 17 demonstrates the comparison between the measured and simulated results. The measurements are in good agreement with simulations. This antenna system provides a cross polarization better than 20 dB in the entire beam steering range.

Finally, Figure 18 demonstrates the scanning range of the antenna system. By increasing the biasing voltage of the varactor diodes from 0 to 20 V, a progressive phase shift is provided and applied to the antennas to steer the beam, resulting in 50° scanning ranges from −25° to 25°. Tables 3 and 4 present a comparison between seven recent antenna systems and our proposed design in terms of gain, phase shifter type, polarization, bandwidth, scan range, and number of antenna elements. As is shown, the proposed system in this paper provides a fairly wide scan range and good gain despite the fact that only two antenna elements are used when compared to other systems, which use four antennas.

The antenna efficiency represents the ratio between the total co-polarized radiated power to the input power accepted by the system. This calculation provides an antenna efficiency of 78.89% at the mid-band frequency of 17 GHz which demonstrates a very good performance of the proposed beam steering system.

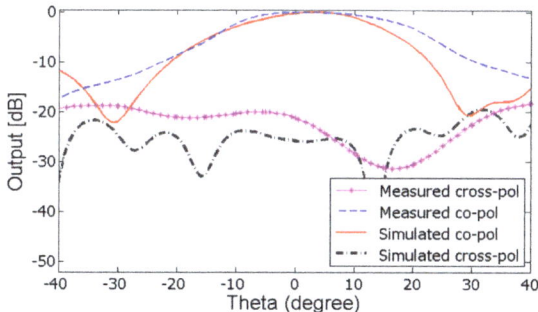

Figure 17. Co-pol and cross pol of the beam steering antenna system when exciting port 2 and applying $V_1 = V_2 = 20$ V at mid-band frequency of 17 GHz.

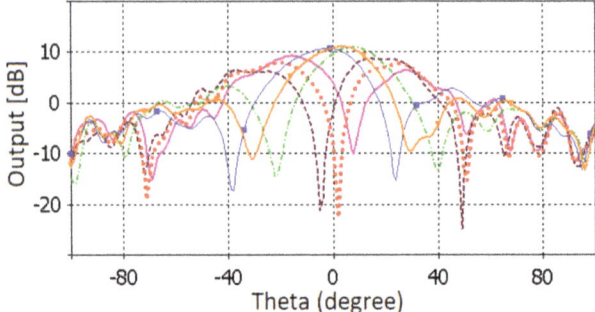

Figure 18. 50° scan range provided by changing the biasing voltage of the varactor diodes from 0 V to 20 V.

Table 3. Comparison of seven recent beam steering antenna systems with our design in terms of phase shifter type, bandwidth, and number of elements.

Antenna System	Number of Antenna Elements	Phase Shifter/Mode	Bandwidth (GHz)
[2]	Four	PIN diode/variable	9.88–10.2
[3]	Four	Nolan matric/fixed	12.25–12.75
[4]	Four	Butler matrix/fixed	27–29
[5]	One	CSRR/fixed	1.98–2.05
[12]	One	PIN diode/fixed	2.8–4.8
[15]	Four	4 × 4 Butler matrix/fixed	23–27
[16]	Four	Metal-insulator-metal Varactor/variable	7–8.4
Our work	Two	Varactor diode/variable	16–18

Table 4. Comparison of seven recent beam steering antenna systems with our design in terms of gain, scan range, and polarization.

Antenna System	Gain [dBi]	Scan Angle (degree)	Polarization
[2]	15	−30 to +30	Linear
[3]	Not measured	Port 1/−5.3 to −15.3 Port 2/28.7 to 38.3 Port 3/6.9 to 16.2 Port 4/−27.7 to −40.7	Linear
[4]	11	−35 to +35	Linear
[5]	Not measured	−51 to +48	Linear
[12]	5.47	Various in different frequencies.	Linear
[15]	8.7–11.7	−25 to +25	Linear
[16]	13	−30 to +30	Linear
Our work	10.7	−25 to +25	Linear

5. Conclusions

A substrate integrated waveguide beam steering antenna system is presented for Ku-band applications. The system uses two reflection-type variable phase shifters, a 3-dB coupler with additional phase transition for creating sum or difference patterns, and two Vivaldi antennas. The entire structure is designed on a single substrate layer which makes it compact, low loss, and low profile. Four varactor diodes are used as electrical steering control elements. Experimental results are in good agreement with simulations, thus verifying the design approach.

Author Contributions: S.S.H. and J.B. conceived the initial designs of the phase shifters, Vivaldi antennas, coupler and 90° phase transition. S.S.H. designed the circuits and experiments, performed the experiments and wrote the first draft of the paper, which both authors finalized.

Funding: This research received no external funding.

Conflicts of Interest: The authors declare no conflicts of interest.

References

1. Ding, Y.; Wu, K. Varactor-tuned substrate integrated waveguide phase shifter. In Proceedings of the 2011 IEEE MTT-S International Microwave Symposium, Baltimore, MA, USA, 5–10 June 2011; pp. 1–4.
2. Errifi, H.; Baghdad, A.; Badri, A.; Sahel, A. Electronically reconfigurable beam steering array antenna using switched line phase shifter. In Proceedings of the 2017 International Conference on Wireless Networks and Mobile Communications (WINCOM), Rabat, Morocco, 1–4 November 2017; pp. 1–6.
3. Djerafi, T.; Fonseca, N.J.G.; Wu, K. Planar Ku-band 4 × 4 Nolen matrix in SIW technology. *IEEE Trans. Microw. Theory Tech.* **2010**, *58*, 259–266. [CrossRef]
4. Khattak, M.K.; Salman Khattak, M.; Rehman, A.; Lee, C.; Han, D.; Park, H.; Kahn, S. A flat, broadband and high gain beam-steering antenna for 5G communication. In Proceedings of the 2017 International Symposium on Antennas and Propagation (ISAP), Phuket, Thailand, 30 October–2 November 2017; pp. 1–2.
5. Cao, W.; Xiang, Y.; Zhang, B.; Liu, A.; Yu, T.; Guo, D. A low-cost compact patch antenna with beam steering based on CSRR-loaded ground. *IEEE Antennas Wirel. Propag. Lett.* **2011**, *10*, 1520–1523. [CrossRef]
6. Bousbia, L.; Ould-Elhassen, M.; Mabrouk, M.; Ghazel, A. New synthesis of tunable filter and phase shifter used for phased array antennas applications. In Proceedings of the 2014 International Conference on Multimedia Computing and Systems (ICMCS), Marrakech, Morocco, 14–16 April 2014; pp. 1348–1353.
7. Ma, S.; Yang, G.-H.; Meng, F.-Y.; Ding, X.-M.; Zhang, K.; Fu, J.-H.; Wu, Q. Electrically tunable array antenna with beam steering from backfire to endfire based on liquid crystal miniaturized phase shifter. In Proceedings of the 2016 IEEE Conference on Electromagnetic Field Computation (CEFC), Miami, FL, USA, 13–16 November 2016; p. 1.
8. Zheng, Y.; Saavedra, C.E. Full 360° vector-sum phase-shifter for microwave system applications. *IEEE Trans. Circuits Syst. I Regul. Pap.* **2009**, *57*, 752–758. [CrossRef]
9. Asoodeh, A.; Mojtaba, A. A full 360° vector-sum phase shifter with very low RMS phase error over a wide bandwidth. *IEEE Trans. Microw. Theory Tech.* **2012**, *60*, 1626–1634. [CrossRef]
10. Peng, Z.; Ran, L.; Li, C. A K-band portable FMCW radar with beamforming array for short-range localization and vital-Doppler targets discrimination. *IEEE Trans. Microw. Theory Tech.* **2017**, *65*, 3443–3452. [CrossRef]
11. Sellal, K.; Talbi, L.; Nedil, M. Design and implementation of a controllable phase shifter using substrate integrated waveguide. *IET Microw. Antennas Propag.* **2012**, *6*, 1090–1094. [CrossRef]
12. Muneer, B.; Qi, Z.; Xu, S. A broadband tunable multilayer substrate integrated waveguide phase shifter. *IEEE Microw. Wirel. Compon. Lett.* **2015**, *25*, 220–222. [CrossRef]
13. Sbarra, E.; Marcaccioli, L.; Gatti, R.V.; Sorrentino, R. Ku-band analogue phase shifter in SIW technology. In Proceedings of the 2009 European Microwave Conference (EuMC), Rome, Italy, 29 September–1 October 2009; pp. 264–267.
14. Formiga Mamedes, D.; Esmaeili, M.; Bornemann, J. K-band substrate integrated waveguide variable phase shifter. In Proceedings of the 2016 10th European Conference on Antennas and Propagation (EuCAP), Davos, Switzerland, 10–15 April 2016; pp. 1–4.
15. Park, Z.; Lin, J. A beam-steering broadband microstrip antenna for noncontact vital sign detection. *IEEE Antennas Wirel. Propag. Lett.* **2011**, *10*, 235–238. [CrossRef]
16. Ala-Laurinaho, J.; Karttunen, A.; Räisänen, A.V. A mm-wave integrated lens antenna for E-band beam steering. In Proceedings of the 2015 9th European Conference on Antennas and Propagation (EuCAP), Lisbon, Portugal, 13–17 April 2015; pp. 1–2.
17. Yao, Y.-L.; Zhang, F.-S.; Zhang, F. A new approach to design circularly polarized beam-steering antenna arrays without phase shift circuits. *IEEE Trans. Antennas Propagat.* **2018**, *66*, 2354–2364. [CrossRef]
18. Bartlett, C.; Salem Hesari, S.; Bornemann, J. End-fire substrate integrated waveguide beam-forming system for 5G applications. In Proceedings of the 2018 18th International Symposium on Antenna Technology and Applied Electromagnetics (ANTEM), Waterloo, ON, Canada, 19–22 August 2018; pp. 1–4.
19. Nikfalazar, M.; Mehmood, A.; Sohrabi, M.; Mikolajek, M.; Wiens, A.; Maune, H.; Kohler, C.; Binder, J.R.; Jakoby, R. Steerable dielectric resonator phased-array antenna based on inkjet-printed tunable phase shifter with BST metal-insulator-metal varactors. *IEEE Antennas Wirel. Propag. Lett.* **2016**, *15*, 877–880. [CrossRef]

20. Kordiboroujeni, Z.; Bornemann, J.; Sieverding, T. Mode-matching design of substrate-integrated waveguide couplers. In Proceedings of the 2012 Asia-Pacific Symposium on Electromagnetic Compatibility, Singapore, 21–24 May 2012; pp. 701–704.
21. Kordiboroujeni, Z.; Bornemann, J. Design of substrate integrated waveguide components using mode-matching techniques. In Proceedings of the 2015 IEEE MTT-S International Conference on Numerical Electromagnetic and Multiphysics Modeling and Optimization (NEMO), Ottawa, ON, Canada, 11–14 August 2015; pp. 1–3.
22. Kordiboroujeni, Z.; Bornemann, J. Designing the width of substrate integrated waveguide structures. *IEEE Microw. Wirel. Compon. Lett.* **2013**, *23*, 518–520. [CrossRef]
23. Skyworks. SMV2201-SMV2205 Series: Surface Mount, 0402 Silicon Hyperabrupt Tuning Varactor Diodes Data Sheet. 2012. Available online: https://www.google.com/url?sa=t&rct=j&q=&esrc=s&source=books&cd=1&ved=2ahUKEwih-pbo7cXkAhVVeXAKHboTAykQFjAAegQIABAC&url=http%3A%2F%2Fwww.skyworksinc.com%2Fuploads%2Fdocuments%2F201953A.pdf&usg=AOvVaw02zMwJrKKzwrIhnLSz5yfD (accessed on 26 July 2019).
24. Locke, L.S.; Bornemann, J.; Claude, S. Substrate integrated waveguide-fed tapered slot antenna with smooth performance characteristics over an ultra-wide bandwidth. *ACES J.* **2013**, *28*, 454–462.
25. Salem Hesari, S.; Bornemann, J. Frequency-selective substrate integrated waveguide front-end system for tracking applications. *IET Microw. Antennas Propag.* **2018**, *12*, 1620–1624. [CrossRef]
26. Balanis, C.A. *Antenna Theory—Analysis and Design*, 3rd ed.; Wiley: New York, NY, USA, 2005.
27. Cheng, Y.; Hong, W.; Wu, K. Novel substrate integrated waveguide fixed phase shifter for 180-degree directional coupler. In Proceedings of the 2007 IEEE/MTT-S International Microwave Symposium, Honolulu, HI, USA, 3–8 June 2007; pp. 189–192.
28. CST. Application Note CST STUDIO SUITE™ 2006B: Antenna Arrays. 2012. Available online: https://www.google.com/url?sa=t&rct=j&q=&esrc=s&source=web&cd=2&ved=2ahUKEwjRlru778XkAhWYHXAKHeFKDusQFjABegQIBhAC&url=https%3A%2F%2Fperso.telecom-paristech.fr%2Fbegaud%2Fintra%2FCST_ANTENNAARRAY.pdf&usg=AOvVaw0UxlbJD3SHsaJahJQ2HgwM (accessed on 26 July 2019).

© 2019 by the authors. Licensee MDPI, Basel, Switzerland. This article is an open access article distributed under the terms and conditions of the Creative Commons Attribution (CC BY) license (http://creativecommons.org/licenses/by/4.0/).

Article

A High-Efficiency K-band MMIC Linear Amplifier Using Diode Compensation

Heng Zhu, Wei Chen *, Jianhua Huang, Zhiyu Wang and Faxin Yu

School of Aeronautics and Astronautics, Zhejiang University, Hangzhou 310027, China; hzhu_zju@zju.edu.cn (H.Z.); oscarhua@zju.edu.cn (J.H.); zywang@zju.edu.cn (Z.W.); fxyu@zju.edu.cn (F.Y.)
* Correspondence: cwydl@zju.edu.cn; Tel.: +86-1876-847-9949

Received: 12 April 2019; Accepted: 29 April 2019; Published: 30 April 2019

Abstract: This paper describes the design and measured performance of a high-efficiency and linearity-enhanced K-band MMIC amplifier fabricated with a 0.15 μm GaAs pHEMT processing technology. The linearization enhancement method utilizing a parallel nonlinear capacitance compensation diode was analyzed and verified. The three-stage MMIC operating at 20–22 GHz obtained an improved third-order intermodulation ratio (IM3) of 20 dBc at a 27 dBm per carrier output power while demonstrating higher than a 27 dB small signal gain and 1-dB compression point output power of 30 dBm with 33% power added efficiency (PAE). The chip dimension was 2.00 mm × 1.40 mm.

Keywords: high-efficiency; K-band; linearity enhancement; power amplifier; GaAs pHEMT; diode compensation

1. Introduction

GaAs MMIC is regarded as the premier power device for the microwave communication system [1] and phase array radar system [2] witnessed in recent decades. However, when facing high peak-to-average ratio (PAR) modulation schemes such as QPSK and OFDM, the nonlinearity of the power amplifier causes spectral reproduction and intermodulation distortion. When multi-signals are amplified within a single channel, the beat between carriers generates amplitude modulation [3].

To meet the linearity requirements in the point-to-point radio or satellite communications which usually operate with a high PAR and inconstant enveloped input signal, conventional designs have to work at a back-off output point compared to their saturated power level. Thus, several techniques have been employed to improve the efficiency in the low power region, such as the linear Doherty design, feed-forward technique, and envelop feedback. Some linear Doherty amplifier [4] and feed-forward designs [5] show a high linearity at an acceptable efficiency; however the complexity and cost of chips are not low. Class-J [6] was also reported to achieve a high linearity in the back-off region. Those technologies usually generate a high circuit complexity. Some literature has also reported on the possibility of inner chip nonlinear compensation methods based on diodes [7]. However, no concrete MMIC design has been proposed.

This paper presents a high-efficiency K-band MMIC linear power amplifier fabricated with a 0.15 μm GaAs pHEMT processing technology. A kind of linearizer circuit of diode nonlinear compensation was accomplished. The Y-parameter matrix method was used to analyze and deduce the dynamic characteristic of the parallel diode and FET network. Both simulation and measurement results show the linearity improvement of the circuit. As a result, the proposed linear amplifier achieved an excellent performance with more than a 1 W output power and 33% power-added efficiency at the 1-dB compression point while maintaining an IM3 better than 20 dBc at an output power of 27 dBm per carrier over a 20–22 GHz band.

2. Nonlinear Analysis and Diode Compensation

Table 1 shows the main parameter of the 0.15 µm GaAs pHEMT process. The equivalent circuit models of pHEMT and a diode are shown in Figure 1.

Table 1. The parameters of the 0.15 µm GaAs pHEMT process.

Parameter	Value	Parameter	Value
V_{TH} (V)	−1.2	f_t (GHz)	85
V_{BDG} (V)	10	Gm_Peak (mS/mm)	495
I_{dmax} (mA/mm)	650	$P_{density}$ (W/mm)	0.8
I_{dss} (mA/mm)	500	C_{MIM} (pF/mm^2)	400

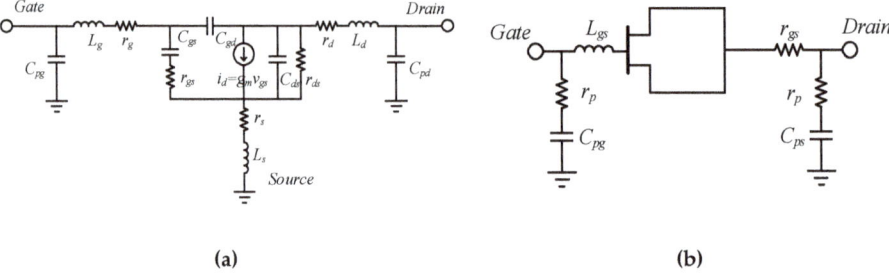

(a)   (b)

Figure 1. (a) The equivalent circuit model of pHEMT; (b) the equivalent circuit model of a diode.

Using a diode depletion capacitance model [8], the values of C_{gs} and C_{gd} can be derived as

$$C_{gs}(v_{gs}) = \frac{C_{gsb}}{\sqrt{1 - \frac{v_{gs}}{v_{bi}}}} \tag{1}$$

$$C_{gd}(v_{gd}) = \frac{C_{gdb}}{\sqrt{1 - \frac{v_{gd}}{v_{bi}}}} \tag{2}$$

The functional model of a single-stage pHEMT amplifier can be represented by Figure 2.

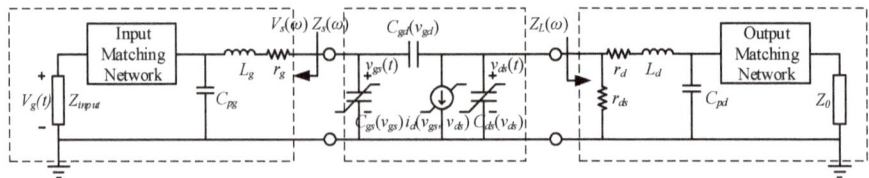

Figure 2. Single-stage pHEMT amplifier model.

In this model, $V_S(\omega)$ and $Z_S(\omega)$ constitute the Thevenin equivalent of the excitation source and the input matching network, and $Z_L(\omega)$ is the load impedance of the pHEMT. Assuming that the pHEMT is excited by a voltage, $v_{gs}(A,t)$:

$$v_{gs}(A,t) = \sum_{0}^{n} V_{gsk}(A,\omega) \cdot \cos(k\omega_0 t) \tag{3}$$

where V_{gsk} represents the Fourier components of v_{gs} and A is the amplitude. By setting the dynamic charge between the gate and source determined by v_{gs} as Q_{gs}, i_{gs} can be derived as

$$i_{gs}(t) = \frac{\partial Q_{gs}[v_{gs}]}{\partial t} = C_{gs}[v_{gs}] \cdot \frac{\partial v_{gs}}{\partial t} \tag{4}$$

Then, we get the frequency domain representation of i_{gs} as

$$I_{gs} = [\frac{1}{2}\sum_{-n}^{n} C_{gsk}(V_{gs}, \omega)] \cdot [\frac{1}{2}\sum_{-n}^{n} jk\omega_0 V_{gsk}] \tag{5}$$

where C_{gsk} is the Fourier components of C_{gs}:

$$C_{gsk} = \frac{\omega}{2\pi} \int_{-\pi/\omega}^{\pi/\omega} C_{gs}[v_{gs}] \cdot e^{-jk\omega_0 t} dt \tag{6}$$

Substituting (R4) into (R3), the fundamental component of i_{gs} would be

$$I_{gs1} = j\omega_0 C_{gs0} V_{gs1} + \frac{1}{2}j2\omega_0 C_{gs1}^* \cdot V_{gs2} - \frac{1}{2}j\omega_0 C_{gs2} \cdot V_{gs1}^* + \ldots \tag{7}$$

As a result, the fundamental $v_{gs}(t)$ voltage V_{gs1} (A) can be derived as

$$V_{gs1} = V_S(\omega) - j\omega Z_s(\omega)\{C_{gs0} \cdot V_{gs1} + C_{gs1}^* \cdot V_{gs2} - \frac{1}{2}C_{gs2} \cdot V_{gs1}^* + \ldots\} \tag{8}$$

and solving (8) leads to

$$V_{gs1} = \frac{V_s(\omega)}{1 + j\omega Z_s(\omega) \cdot C_{gs0}} \tag{9}$$

The term involving C_{gs2} and V_{gs2} is neglected as it is significantly smaller than the others. Incorporating the feedback capacitor C_{gd} into (9), we get

$$V_{gs1} = \frac{V_s(\omega)}{1 + j\omega Z_s(\omega) \cdot [C_{gs0} + C_{gd}(1 - A_v)]} \tag{10}$$

(10) reveals that C_{gs} and C_{gd} influence the phase of fundamental $v_{gs}(t)$, which leads to AM/PM distortion. Using a similar method, we can derive the fundamental $v_{ds}(t)$ voltage V_{ds1} (A) as

$$V_{ds1}(A) \approx \frac{Z_L(\omega) \cdot I_{d1}(A)}{1 + j\omega Z_L(\omega) \cdot [C_{ds0}(A) + C_{gd}(\frac{A_v - 1}{A_v})]} \tag{11}$$

The term involving C_{ds2} is neglected as it is significantly smaller than the others. Since the denominator of (11) is approximately equal to unity [9], as a result, there is no significant impact on AM/PM brought about by C_{ds}. From (10) and (11), it can be derived that the main contributors to the AM/PM distortion are the variations of C_{gs} and C_{gd}. According to the barrier capacitance effect, the value of C_{gs} is usually much bigger than that of C_{gd}. Therefore, the nonlinear characteristic of pHEMT is mainly contributed by C_{gs}, which approximately equals C_{in}. As shown in Figure 3, a Schottky diode is parallel in the gate and the drain of the output stage FET to compensate for the nonlinearity of C_{gs}.

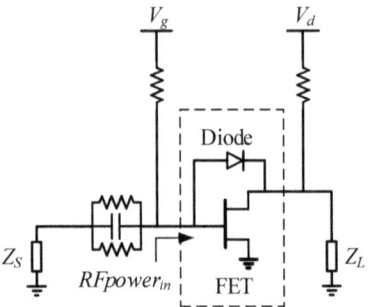

Figure 3. Paralleled Shottky diode linearizer.

The voltage between the reverse diode is

$$v_{diode} = v_{ds} - v_{gs} \tag{12}$$

The two-port Y-parameter matrix can be utilized to analyse the parallel network composed of the FET and diode. The matrix form of FET is as follows:

$$\begin{bmatrix} i_g \\ i_d \end{bmatrix} = \begin{bmatrix} y_{11} & y_{12} \\ y_{21} & y_{22} \end{bmatrix} \begin{bmatrix} v_{gs} \\ v_{ds} \end{bmatrix} \tag{13}$$

where

$$y_{11} = \frac{j\omega C_{gs}}{1 + j\omega C_{gs}r_{gs}} + j\omega C_{gd} \tag{14}$$

$$y_{12} = -j\omega C_{gd} \tag{15}$$

$$y_{21} = \frac{g_m}{1 + j\omega C_{gs}R_{gs}} - j\omega C_{gd} \tag{16}$$

$$y_{22} = \frac{1}{r_{ds}} + j\omega(C_{ds} + C_{gs}) \tag{17}$$

Incorporating the diode barrier capacitance into the matrix,

$$y_{11}' = \frac{j\omega C_{gs}}{1 + j\omega C_{gs}r_{gs}} + j\omega(C_{gd} + C_{diode}) \tag{18}$$

$$C_{in}' = -\frac{1}{\omega \text{Im}\left(\frac{1}{y_{11}'}\right)} \tag{19}$$

C_{in}' represents the input capacitance of the network. According to the previous analysis, the state of input capacitance mainly determines the nonlinear characteristics of the network. Thus, it is possible to perform distortion compensation utilizing the high-frequency C-V characteristic of the diode. As $v_{diode} < 0$, C_{diode} is dominated by depletion layer capacitance C_j, where

$$C_j = \frac{C_{j0}}{\sqrt{1 - \frac{v_{diode}}{\Phi_B}}} \tag{20}$$

where Φ_B is the junction built-in potential, so

$$C_{in}' = \frac{[\omega C_{gs}R_{gs}(C_{gd} + C_j)]^2 + (C_{gs} + C_{gd} + C_j)^2}{(\omega C_{gs}R_{gs})^2(C_{gd} + C_j) - (C_{gs} + C_{gd} + C_j)} \tag{21}$$

As the input RF power increases, the electric field within the FET channel is enhanced, which makes the swing of both v_{gs} and v_{ds} become higher. However, the negative swing of v_{ds} is limited by the knee voltage (V_{knee}) and the positive swing of v_{gs} is limited by the diffusion barrier voltage (V_{diff}). As a result, $\bar{v}_{ds}$ grows higher while $\bar{v}_{gs}$ grows lower. Consequently, $\bar{v}_{diode}$ (3) becomes higher while $\bar{v}_{gs}$ becomes lower. The voltage waveform versus input RF power is shown in Figure 4.

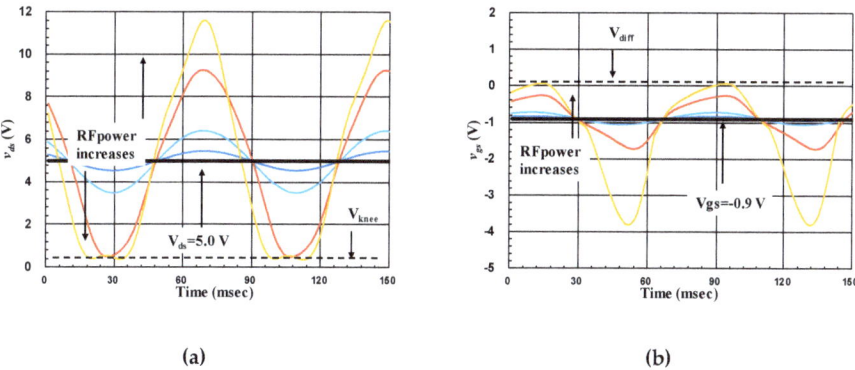

Figure 4. (a) Voltage waveform of v_{ds}; (b) voltage waveform of v_{gs}.

Therefore, substituting (1), (2), and (20) into (21) and combining the voltage analysis above, the characteristics of input capacitance can be shown as the curve in Figure 5.

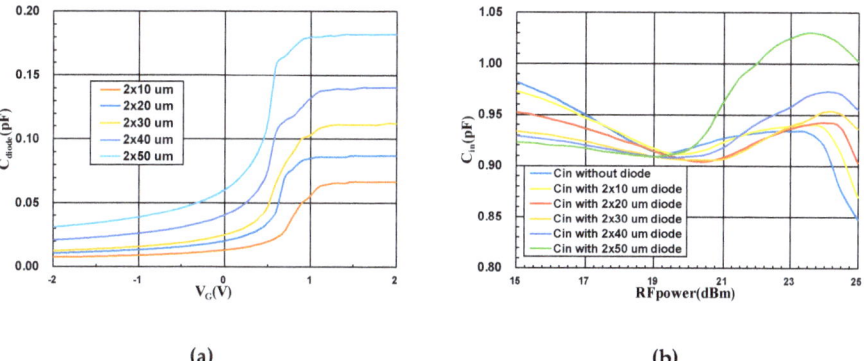

Figure 5. (a) C-V characteristics of the two-finger diode; (b) input capacitance of the network versus RF power.

The principles for the choice of diode periphery are as follows.

1. The size of the diode should be large enough to compensate for the input capacitance of the pHEMT as the input power increases;
2. The barrier capacitance of the diode should not be too large to over change the input characteristics of the HMET and generate excessive feedback between drain and gate.

Figure 6 shows that as the size of diode increases, the additional feedback becomes stronger. As a result, the MaxGain of the network decreases, which leads to lower linear gain. Therefore, combined with the C-V characteristics of the diode shown in Figure 5a, we gradually adjusted the diode size so that it could effectively compensate for the input capacitance of pHEMT while reducing the effect on the gain. The C-V characteristics of the selected 2 × 20 µm diode and the input capacitance of the network versus RF power are shown in Figure 7.

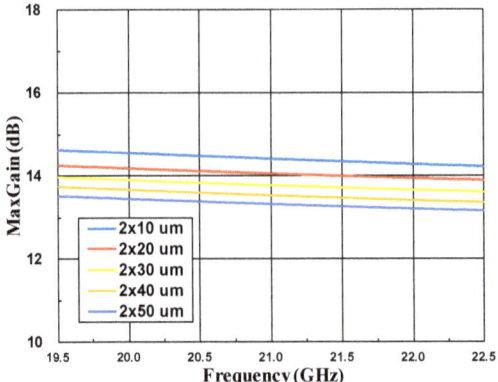

Figure 6. The influence of the diode size on the MaxGain of the network.

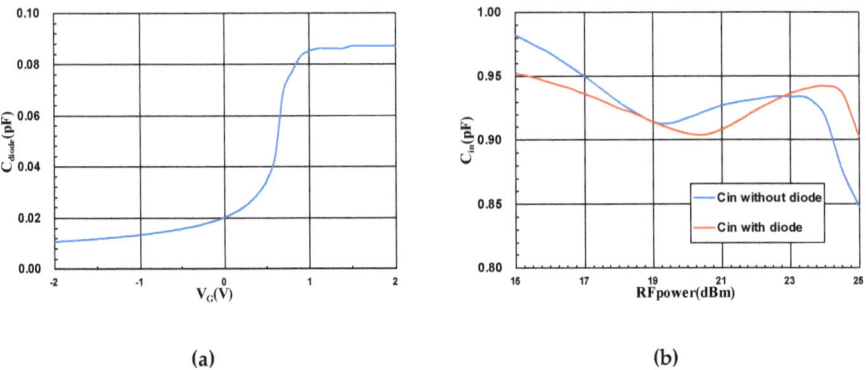

Figure 7. (a) C-V characteristics of the 2 × 20 µm diode; (b) input capacitance of the network versus RF power (Behavior of 2 × 20 µm diode and 8 × 100 µm FET).

The parasitic parameter value of the selected pHEMT and diode is shown in Tables 2 and 3, respectively.

Table 2. The parasitic parameter value of the 8 × 100 µm pHEMT

Parameter	Value	Parameter	Value
r_g (Ω)	1.5	L_d (pH)	8.5
r_d (Ω)	1.2	L_s (pH)	0.7
r_s (Ω)	0.2	C_{pg} (fF)	9.5
L_g (pH)	9.5	C_{pd} (fF)	17.5

Table 3. The parasitic parameter value of the 2 × 20 µm diode

Parameter	Value
r_{gs} (Ω)	41.5
C_{pg} (fF)	89.0
C_{ps} (fF)	33.5
L_{gs} (pH)	55.0

Converting (7) to the S-parameter matrix:

$$\begin{bmatrix} b_1 \\ b_2 \end{bmatrix} = \begin{bmatrix} S_{11} & S_{12} \\ S_{21} & S_{22} \end{bmatrix} \begin{bmatrix} a_1 \\ a_2 \end{bmatrix} \quad (22)$$

where

$$S_{11} = \frac{(1 - Z_0 y_{11})(1 + Z_0 y_{22}) + y_{12} y_{21} Z_0^2}{\Psi} \quad (23)$$

$$S_{12} = \frac{-2Y_{12} Z_0}{\Psi} \quad (24)$$

$$S_{21} = \frac{-2Y_{21} Z_0}{\Psi} \quad (25)$$

$$S_{22} = \frac{(1 + Z_0 y_{11})(1 - Z_0 y_{22}) + y_{12} y_{21} Z_0^2}{\Psi} \quad (26)$$

$$\Psi = (1 + Z_0 y_{11})(1 + Z_0 y_{22}) - y_{12} y_{21} Z_0^2 \quad (27)$$

Using the S-parameter to derive the stability factor of the network:

$$k = \frac{1 - |S_{11}|^2 - |S_{22}|^2 + |\Delta|^2}{2|S_{12}||S_{21}|} \quad (28)$$

$$\Delta = S_{11} S_{22} - S_{12} S_{21} \quad (29)$$

The factor $k > 1$ is a necessary and sufficient condition for network stability. Since the diode compensation method generates additional feedback between the gate and drain, the stability of the network is enhanced compared with single FET. The conclusion can be verified by calculation or simulation. The simulation result of the stability factor is shown in Figure 8.

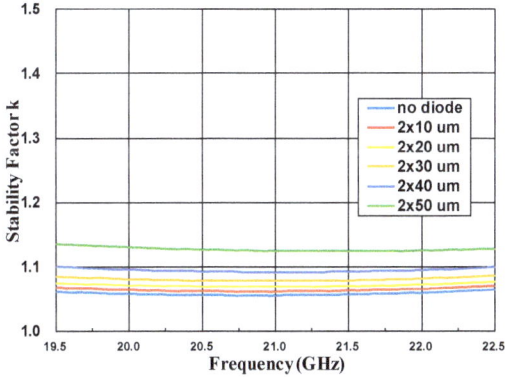

Figure 8. The influence of the diode size on the stability factor of the network.

Consequently, because of the nonlinear effect of the diodes, the input capacitance of the network is compensated for as the input power increases, thus avoiding the degradation of linearity. As the diode is operated in an inverted connection, there is nearly no additional current or power loss in the circuit.

3. Circuit Design

The K-band MMIC linear amplifier is composed of three-stage FETs fabricated with a 0.15 µm gate length AlGaAs/GaAs pHEMT technology. The process exhibits a gate-drain breakdown voltage of 16.5 V and a cutoff frequency (f_T) of 90 GHz. The FETs for the power stage are 8 × 100 µm. Source-pull and load-pull simulations using a large signal model at a center frequency were taken to determine the optimal input and output impedances that lead to a higher output power and efficiency. The output matching network combining two FETs matches the fundamental load impedance based on the load-pull simulation. The interstage and input matching networks are designed to match the conjugated impedance and were optimized for low loss. The circuit of the amplifier configuration is shown in Figure 9. The simulation result of the amplifier is shown in Figures 10 and 11.

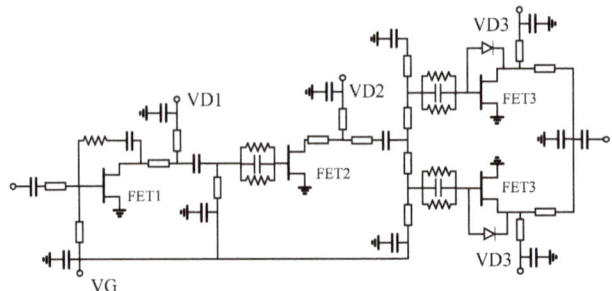

Figure 9. Schematic of the MMIC linear amplifier.

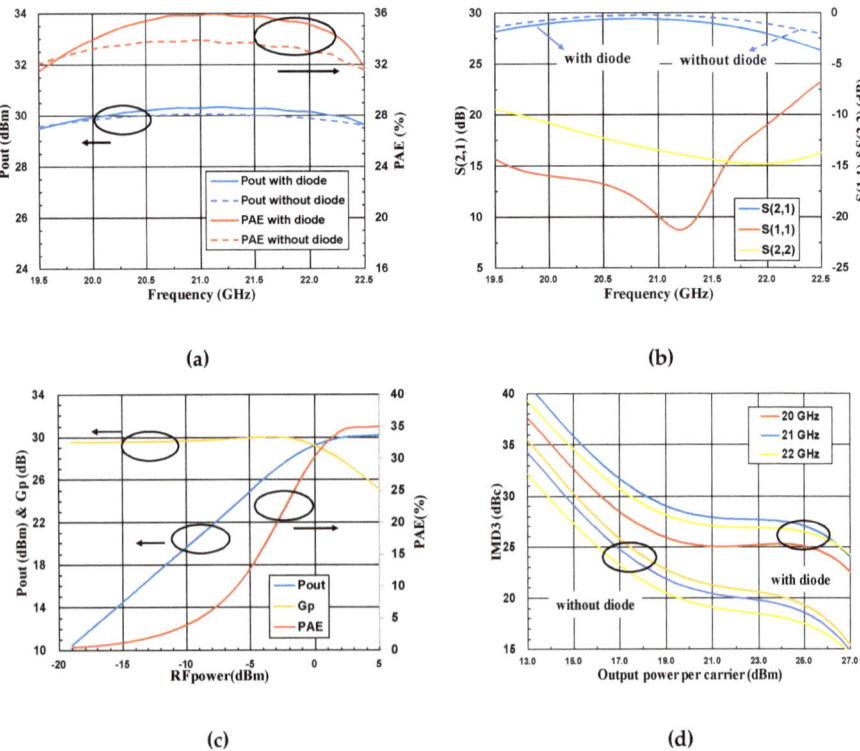

Figure 10. (a) Simulated output power and PAE versus frequency at a 3 dBm input power; (b) simulated S-parameter versus frequency; (c) simulated output power, gain, and PAE versus input power at a 21 GHz frequency; (d) simulated IMD3 and PAE versus output power.

Figures 10 and 11 show the improvement of the amplifier linearity performance generated by the parallel diode circuit design.

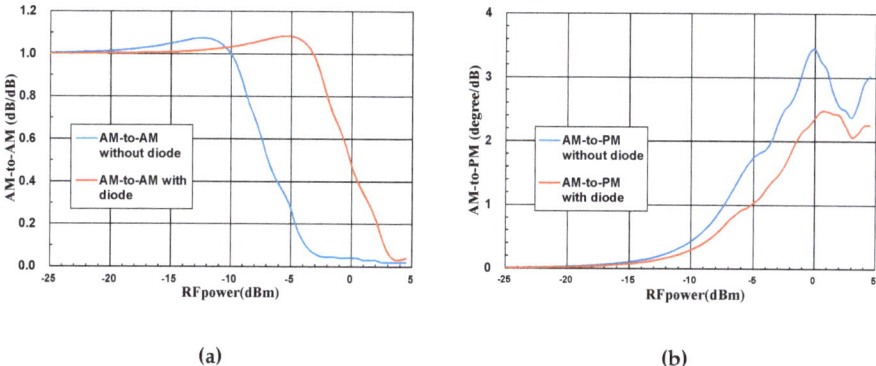

Figure 11. (a) Simulated AM-to-AM versus input power at 21 GHz; (b) simulated AM-to-PM versus input power at 21 GHz.

4. Measurement Results

The fabricated three-stage K-band linear amplifier MMIC is shown in Figure 12. The MMIC size is as small as 2.00 × 1.40 mm² with a GaAs substrate thickness of 50 μm. Linear S-parameter and large signal measurements were performed at a drain voltage of 5 V and gate voltage of −0.8V on a wafer. Figure 13 shows the measured S-parameter of the MMIC from 19.5 GHz to 22.5 GHz. The design has achieved higher than a 26 dB small signal gain and better than a 10 dB input return loss. Figure 14 illustrates the measured output power and PAE for the amplifier at fixed input drive levels of +3.5 dBm under a continuous wave (CW). At the nominal supply of VD = 5 V and VG = −0.8 V, the amplifier demonstrates higher than 30 dBm P1dB with 33–35% PAE over a 20–22 GHz frequency. Very flat power and gain characteristics were achieved for this design. Further PAE improvement should be possible with a more precise power match by the result of a load-pull test.

Two-tone measurement of the amplifier has been performed with 10 MHz tone spacing under the drain voltage of 5 V and gate voltage of −0.8 V. The intermodulation distortion was measured on a wafer with 10 MHz two-tone spacing. The two carrier sources were connected to the chip through a power combiner with an isolator in order to reduce the intermodulation contributions of the setup. As shown in Figure 15, IMD3 better than 20 dBc and PAE higher than 33% were measured at an output power of 27 dBm per carrier over a 20–22 GHz band. Compared to the design without an inner paralleled diode configuration, there is more than a 5dB improvement of IMD3. Table 4 summarizes the performance comparison of this work with other published linear amplifiers working in close frequency ranges. The characteristics of the amplifier, including power, gain, efficiency, and IMD3, are better than the previously reported ones.

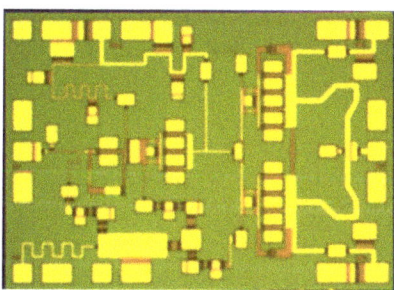

Figure 12. Photograph of the fabricated K-band linear amplifier.

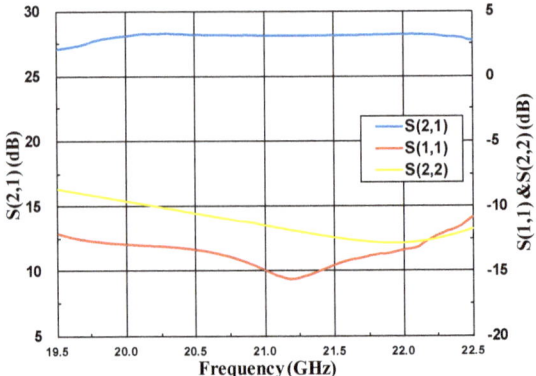

Figure 13. Measured small-signal performance of the amplifier under the bias of VD = 5 V and VG = −0.8 V at 25 Celsius.

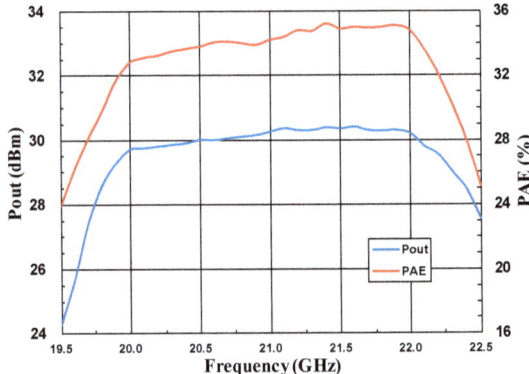

Figure 14. Measured output power and PAE at 1 dB compression.

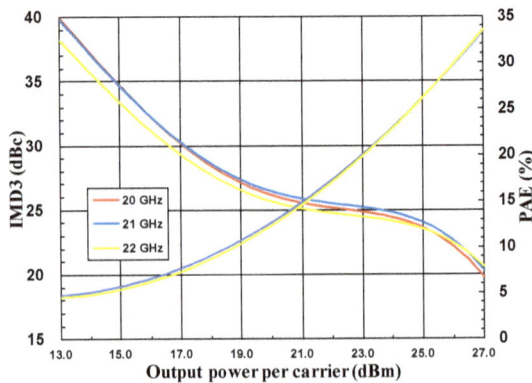

Figure 15. Measured IMD3 and PAE versus output power under a two-tone test with different frequencies.

Table 4. Performance comparison of linear amplifiers.

Reference	Process	Frequency (GHz)	P_{-1dB} (dBm)	PAE (%)	Gain (dB)	IMD3 (dBc)	Size (mm^2)
[10]	0.15 µm GaAs	20–23	31	24	9.5	-	-
[11]	0.25 µm GaAs	22.8–23.5	28	15.3	25	-	2.52
[12]	0.25 µm GaAs	18–27	31.4	27	14	18	3.91
[13]	0.15 µm GaAs	17–20	28	30	21	14	2.96
[14]	0.15 µm GaAs	24–28	28	21	21	15	4.50
This work	0.15 µm GaAs	20–22	30	34	27	20	2.80

5. Conclusions

In this paper, a three-stage K-band 20–22 GHz high-efficiency linear MMIC power amplifier using 0.15 µm GaAs pHEMT technology is reported. The design utilizes an optimum paralleled diode circuit for inner chip linear compensation. The MMIC generates an output power of 30 dBm and PAE of 33% at −1 dB gain compression under CW operation and delivers a lower than −20 dBc IMD3 performance with 10 MHz tone spacing. The MMIC has a smaller size and more than a 5dB improvement of IMD3 was observed.

Author Contributions: Methodology, H.Z.; Software, J.H.; Validation, J.H.; Data Curation, H.Z.; Writing: Original Draft Preparation, H.Z.; Writing: Review and Editing, W.C.; Supervision, W.C.; Project Administration, Z.W.; Funding Acquisition, F.Y.

Funding: This research was funded by the National Natural Science Foundation of China grant number 61604128 and the Fundamental Research Funds for the Central Universities grant number 2017QN81002.

Acknowledgments: The authors would like to thank the Institute of Aerospace Electronics Engineering of Zhejiang University for providing the research platform and technical support.

Conflicts of Interest: The authors declare no conflict of interest.

References

1. Camarchia, V.; Guerrieri, S.D.; Ghione, G.; Pirola, M.; Quaglia, R.; Rubio, J.M.; Loran, B.; Palomba, F.; Sivverini, G. A K-band GaAs MMIC Doherty power amplifier for point-to-point microwave backhaul applications. In Proceedings of the 2014 International Workshop on Integrated Nonlinear Microwave and Millimetre-wave Circuits (INMMiC), Leuven, Belgium, 2–4 April 2014. [CrossRef]
2. Lane, A.; Jenkins, J.; Green, C.; Myers, F. S and C band GaAs multifunction MMICs for phased array radar. In Proceedings of the 11th Annual Gallium Arsenide Integrated Circuit (GaAs IC) Symposium, San Diego, CA, USA, 22–25 October 1989; pp. 259–262. [CrossRef]
3. Simons, K.A. The decibel relationships between amplifier distortion products. *Proc. IEEE* **1970**, *58*, 1071–1086. [CrossRef]
4. Cho, K.-J.; Kim, W.-J.; Kim, J.-H.; Stapleton, S.P. Linearity optimization of a high power Doherty amplifier based on post-distortion compensation. *IEEE Microw. Wireless Compon. Lett.* **2005**, *15*, 748–750. [CrossRef]
5. Roy, M.K. Distortion cancellation performance of miniature delay filters for feed-forward linear power amplifiers. *IEEE Trans. Ultrason. Ferroelectr. Freq. Control* **2002**, *49*, 1592–1595. [CrossRef]
6. Wright, P.; Lees, J.; Benedikt, J.; Tasker, P.J.; Cripps, S.C. A methodology for realizing high efficiency class-J in a linear and broadband PA. *IEEE Trans. Microw. Theory Techn.* **2009**, *57*, 3196–3204. [CrossRef]
7. Mrunal, A.K.; Shirasgaonkar, M.; Patrikar, R.M. Power amplifier linearization using a diode. In Proceedings of the MELECON 2006–2006 IEEE Mediterranean Electrotechnical Conference, Malaga, Spain, 16–19 May 2006; pp. 173–176. [CrossRef]
8. Wren, M.; Brazil, T.J. The effect of the gate Schottky diode on pHEMT power amplifier performance. In Proceedings of the 2003 High Frequency Postgraduate Student Colloquium, Belfast, Ireland, 8–9 September 2003; pp. 52–55. [CrossRef]
9. Nunes, L.C.; Cabral, P.M.; Pedro, J.C. A Physical Model of Power Amplifiers AM/AM and AM/PM Distortions and Their Internal Relationship. In Proceedings of the 2013 IEEE MTT-S International Microwave Symposium Digest (MTT), Seattle, WA, USA, 2–7 June 2013. [CrossRef]

10. Fersch, T.; Quaglia, R.; Pirola, M.; Camarchia, V.; Ramella, C.; Khoshkholgh, A.J.; Ghione, G.; Weigel, R. Stacked GaAs pHEMTs: design of a K-band power amplifier and experimental characterization of mismatch effects. In Proceedings of the 2015 IEEE MTT-S International Microwave Symposium, Phoenix, AZ, USA, 17–22 May 2015; pp. 1–4. [CrossRef]
11. Koo, B.; Park, C.; Lee, K.A.; Chun, J.H.; Hong, S. A 28-dBm pHEMT Power Amplifier Using Voltage Combiner for K-Band Applications. In Proceedings of the 38th European Microwave Conference, Amsterdam, Netherlands, 27–31 October 2008. [CrossRef]
12. Brown, S.A.; Carroll, J.M. Compact, 1 Watt, power amplifier MMICs for K-band applications. In Proceedings of the GaAs IC Symposium, Seattle, WA, USA, 5–8 November 2000.
13. Bessemoulin, A.; Mcculloch, M.G.; Alexander, A.; Mccann, D.; Mahon, S.J.; Harvey, J.T. Compact K-band watt-level GaAsPHEMT power amplifier MMIC with integrated ESD protection. In Proceedings of the 2006 European Microwave Integrated Circuits Conference, Manchester, UK, 10–13 September 2006.
14. Wang, K.; Yan, Y.; Liang, X. A K-band power amplifier in a 0.15-um GaAs pHEMT process. In Proceedings of the 2018 IEEE MTT-S International Wireless Symposium (IWS), Chengdu, China, 6–10 May 2018; pp. 1–3.

© 2019 by the authors. Licensee MDPI, Basel, Switzerland. This article is an open access article distributed under the terms and conditions of the Creative Commons Attribution (CC BY) license (http://creativecommons.org/licenses/by/4.0/).

Article

Design of Broadband W-Band Waveguide Package and Application to Low Noise Amplifier Module

Jihoon Doo, Woojin Park, Wonseok Choe and Jinho Jeong *

Department of Electronic Engineering, Sogang University, Seoul 04107, Korea; dojh1000@sogang.ac.kr (J.D.); woojin8931@naver.com (W.P.); choe0039@sogang.ac.kr (W.C.)
* Correspondence: jjeong@sogang.ac.kr; Tel.: +82-2-705-8934

Received: 27 April 2019; Accepted: 8 May 2019; Published: 10 May 2019

Abstract: In this paper, the broadband millimeter-wave waveguide package, which can cover the entire W-band (75–110 GHz) is presented and applied to build a low noise amplifier module. For this purpose, a broadband waveguide-to-microstrip transition was designed using an extended E-plane probe in a low-loss and thin dielectric substrate. The end of the probe substrate was firmly fixed on to the waveguide wall in order to minimize the performance degradation caused by the probable bending of the substrate. In addition, we predicted and analyzed in-band resonances by the simulations that are caused by the empty spaces in the waveguide package to accommodate integrated circuits (ICs) and external bias circuits. These resonances are removed by designing an asymmetrical bias space structure with a radiation boundary at an external bias connection plane. The bond-wires, which are used to connect the ICs with the transition, can generate impedance mismatches and limit the bandwidth performance of the waveguide package. Their effect is carefully compensated for by designing the broadband two-section matching circuits in the transition substrate. Finally, the broadband waveguide package is designed using a commercial three-dimensional electromagnetic structure simulator and applied to build a W-band low noise amplifier module. The measurement of the back-to-back connected waveguide-to-microstrip transition including the empty spaces for the ICs and bias circuits showed the insertion loss less than 3.5 dB and return loss higher than 13.3 dB across the entire W-band without any in-band resonances. The measured insertion loss includes the losses of 8.7 mm-long microstrip line and 41.8 mm-long waveguide section. The designed waveguide package was utilized to build the low noise amplifier module that had a measured gain greater than 14.9 dB from 75 GHz to 105 GHz (>12.9 dB at the entire W-band) and noise figure less than 4.4 dB from 93.5 GHz to 94.5 GHz.

Keywords: low noise amplifier; millimeter-wave; package; transition; waveguide

---

1. Introduction

Millimeter-wave frequencies including W-band (75–110 GHz) have short wavelengths and broad allocated bandwidth, which are beneficial for high resolution radars and high speed wireless communications [1–5]. In the millimeter-wave frequency range, a rectangular waveguide is generally adopted as a main transmission line instead of a coaxial cable due to the lower loss and easier manufacturability [6]. Therefore, millimeter-wave circuits including integrated circuits (ICs) should be packaged in the rectangular waveguide. This waveguide package needs to be designed to have broadband and low-loss performance. However, there are several design issues to be solved for high performance millimeter-wave waveguide packages [6,7].

The rectangular waveguide has a dominant electromagnetic (EM) field mode of a transverse electric (TE_{10}), so that broadband waveguide-to-microstrip transition is an essential component in order to package the ICs in a microstrip [8–19]. In addition, the empty spaces inside the metallic waveguide are required to accommodate ICs and their bias circuits, and they form metallic cavities that can generate in-band resonances, seriously degrading the performance of the waveguide packages [11,12]. Finally, bond-wires, which are generally used to electrically connect the ICs to the transition, can produce impedance mismatches and limit the bandwidth performance, especially at millimeter-wave frequencies. Therefore, their effect should be accurately modeled and corrected not to degrade the performance when the IC is packaged in the waveguide [20].

In this work, we present the broadband waveguide package at W-band, by designing a broadband low-loss waveguide-to-microstrip transition using an extended E-plane probe, analyzing and removing the in-band resonances caused by the cavities in the waveguide package, and designing broadband bond-wire compensation circuits. These in-band resonance problems and the bond-wire compensation circuits have not been deeply treated in the previous publications on W-band waveguide modules [21,22]. The waveguide package is designed and applied to build the broadband low noise amplifier (LNA) module. A three-dimensional (3-D) EM simulator is used in the analysis and design of the W-band waveguide package. The designed transition and LNA module are both fabricated to show the resonance-free broadband performance covering the entire W-band.

2. Design of Broadband W-band Waveguide Package

2.1. Waveguide-to-Microstrip Transition Using Extended E-Plane Probe

There has been extensive research in developing millimeter-wave waveguide transitions using E-plane probes [8,9], dipole antenna [10–12], antipodal finlines [13,14], and substrate integrated waveguides [15]. In this work, the E-plane probe transition was adopted because of its small size and low-loss performance, as shown in Figure 1. The transition substrate is placed in the central E-plane of waveguide and supported on the channel. The square metallic patch on the substrate captures and transforms EM fields in the TE_{10} mode of the waveguide to the quasi-transverse electromagnetic (TEM) mode of the microstrip line. The high impedance line is used to match the impedances of the probe to that of the microstrip line. In order to minimize the reflection from air-filled waveguide, very thin substrate with low dielectric constant (127 μm-thick TLY-5 with ε_r = 2.2 by Taconic) is used as a transition substrate, considering a small internal size of the standard W-band waveguide (WR-10: 2.54 mm × 1.27 mm). This very thin and soft substrate, however, can be easily bent by a small force during the packaging process, which can lead to serious performance degradation and poor repeatability. In order to alleviate these problems, the substrate in the probe section was extended to the other side of the waveguide and fixed into the slit [16].

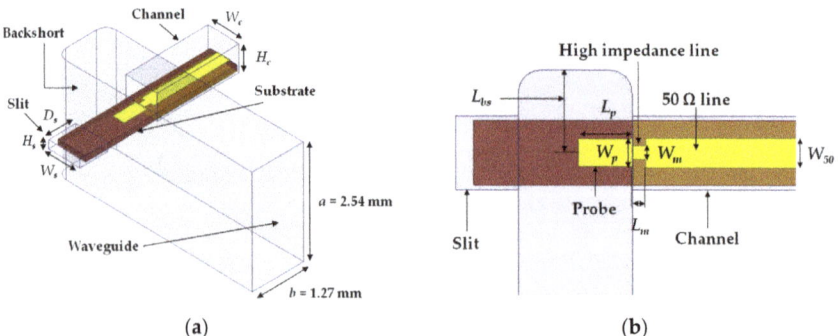

Figure 1. Extended E-plane probe transition: (a) 3-D view and (b) cross-sectional view.

The dimensions of the transition were determined by the commercial 3-D structure simulator (HFSS by Ansys in Canonsburg, PA, USA) and listed in Table 1. At first, the transition was designed without the slit region. Then, the slit was inserted in the simulation and the dimensions were slightly modified. The slit exhibits the marginal effect on the transition performance. Figure 2 shows the simulated performance of the designed transition. The insertion loss ($-10\log|S_{21}|^2$) was less than 0.6 dB and return loss ($-10\log|S_{11}|^2$) was higher than 21.6 dB across the whole W-band.

Table 1. Dimensions of the designed transition (in mm).

W_p	L_p	W_m	L_m	W_c	H_c	W_s	H_s	D_s	L_{bs}	W_{50}
0.3	0.6	0.15	0.15	0.8	0.6	0.8	0.2	0.7	0.9	0.32

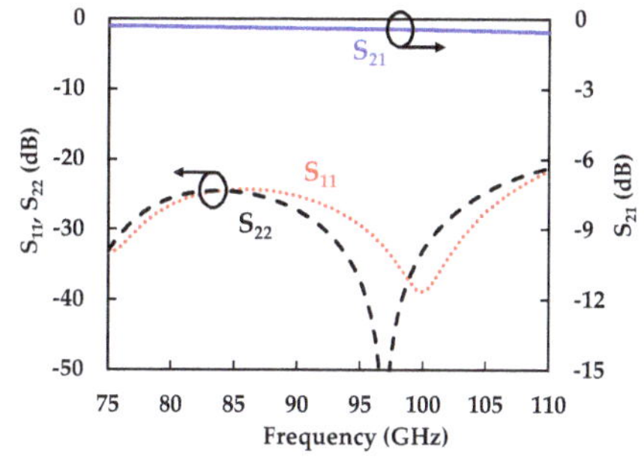

Figure 2. Simulated S-parameters of the designed E-plane probe transition.

2.2. Resonances-Free Waveguide Package

The waveguide package with the transitions necessarily requires several empty spaces to accommodate the circuits, including ICs and direct current (DC) bias circuits. Figure 3 shows the waveguide package with the transitions to package the W-band LNA IC (size = 2 mm × 3 mm). The LNA IC will sit in the place of the microstrip line in the middle. There are several empty spaces for the circuits to supply DC bias to the LNA IC from external power supplies. Unfortunately, these empty spaces can form metallic cavities and generate parasitic resonances that can be observed in the EM simulations. Note that this waveguide package is all closed by the metal except for the waveguide input and output ports. The plane A can be either open or closed for the external bias supply connection.

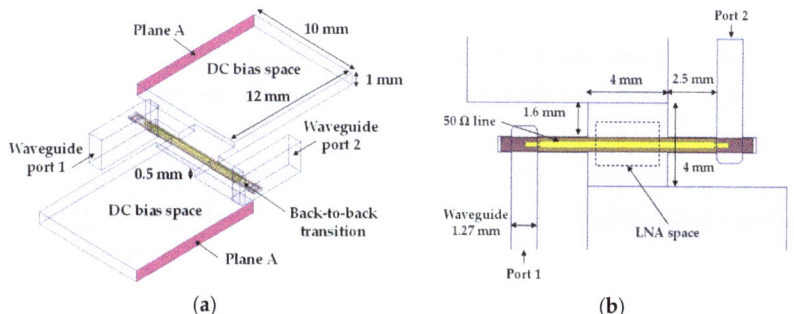

Figure 3. Waveguide package for the low noise amplifier integrated circuit (LNA IC): (a) 3-D view and (b) cross-sectional view.

Figure 4a shows the simulated performance of the waveguide package of Figure 3 (back-to-back transition with several empty spaces) with the plane A closed. There are many resonances at in-band frequencies at which insertion and return losses are seriously degraded. Figure 4c shows the electric field distribution at one of resonance frequencies (86.1 GHz). It can be stated from this figure that this resonance was generated in the empty spaces for the IC and bias circuits, which work as metallic cavities.

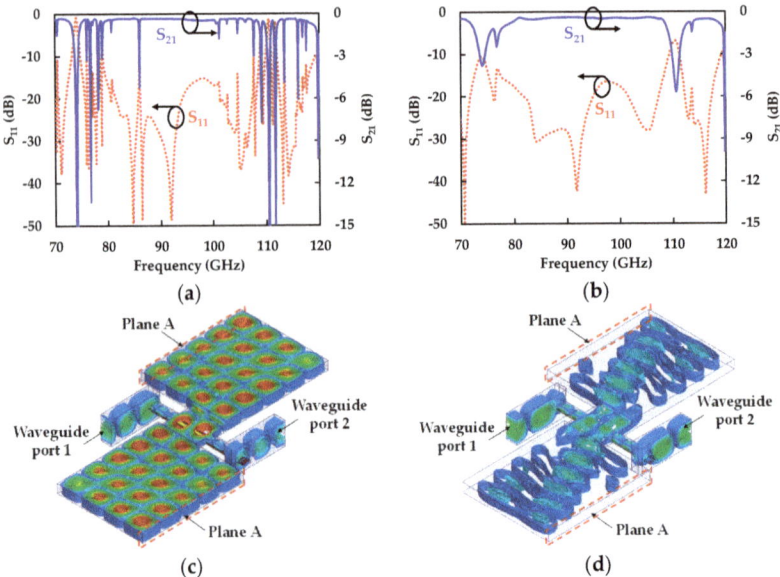

Figure 4. Simulation of the waveguide package shown in Figure 3. (**a**) S-parameters with a metal boundary at plane A. (**b**) S-parameters with a radiation boundary at plane A. (**c**) Electric field distribution at 86.1 GHz (with a metal boundary at plane A). (**d**) Electric field distribution at 76.0 GHz (with a radiation boundary at plane A).

These resonances can be partly suppressed by applying a radiation boundary to the plane A instead of the metal boundary, as shown in Figure 4b. However, there still exists a few in-band resonances. At 76.0 GHz, the electric field distribution is drawn in Figure 4d, showing that the strong electric field was invoked in the IC space and leaked into the bias spaces. Both Figure 4c,d demonstrated that the resonances could be produced in the LNA space. Note that this space was enclosed by the metallic walls and the open boundaries as shown in Figure 4d, forming the rectangular cavity-like structure. The fields in this structure were excited by the microstrip line.

In order to get rid of all the in-band resonances, the LNA space was designed to have an asymmetrical structure as shown in Figure 5a, deforming the rectangular cavity-like structure. It can hinder an electric field from being concentrated in the IC space and thus suppress the resonances. Figure 5b shows the simulated S-parameters of the waveguide package with an asymmetrical structure (with the radiation boundary at the plane A), showing insertion loss less than 1.3 dB and return loss higher than 18.7 dB across the full W-band without any in-band resonances. Note that the insertion loss is that of the back-to-back transitions including the 8.7 mm-long microstrip line.

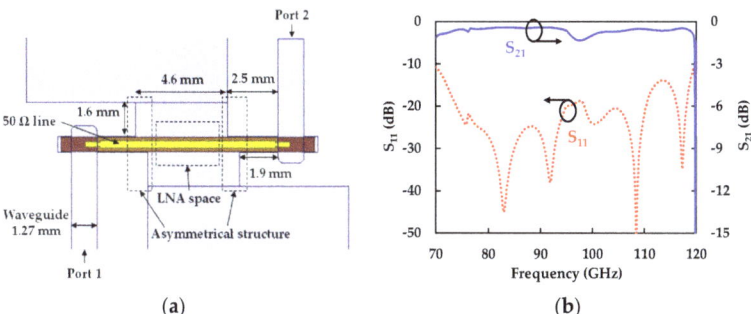

Figure 5. Waveguide package with asymmetrical structure. (a) Cross-sectional view. (b) Simulated S-parameters.

2.3. Compensation of Bond-Wire Effect

In general, bond-wires are used to electrically connect the IC with the microstrip line in the transition substrate. Their effect can be modeled using series inductances and shunt capacitances that can produce serious impedance mismatches, especially at millimeter-wave frequencies [20]. Figure 6a shows the bond-wire connection between 127 µm-thick substrate and 100 µm-thick GaAs substrate ($\varepsilon_r = 12.9$) with a 200 µm distance. A gold wire with a diameter of 25 µm was placed 40 µm above the copper microstrip line in the TLY-5 substrate. Figure 6c shows the simulated S-parameters at W-band showing the increased insertion loss and poor return loss by the bond-wire.

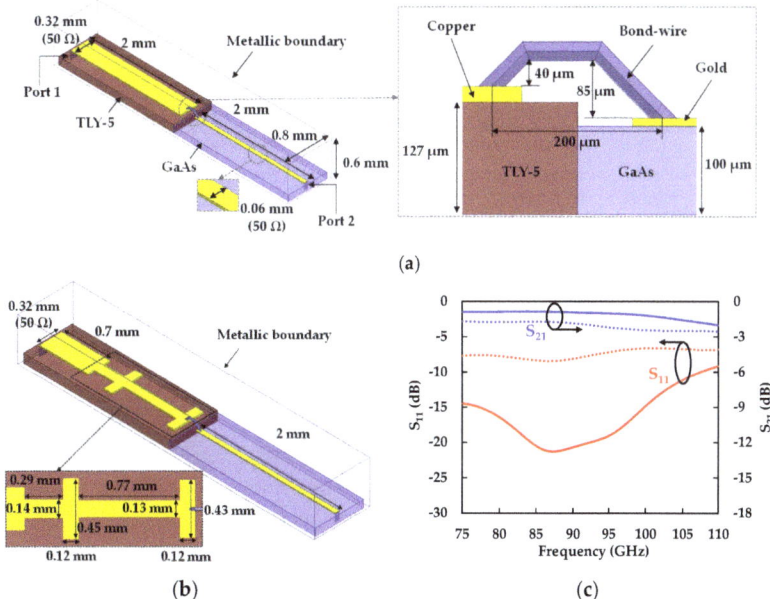

Figure 6. Bond-wire simulation. (a) Basic structure of bond-wire connection. (b) Bond-wire compensation circuit. (c) Simulated S-parameters with compensation circuit (solid) and without compensation circuit (dotted).

In order to reduce impedance mismatches by the bond-wire, a broadband compensation circuit was designed using two-section matching network as shown in Figure 6b. The simulation shows that the bond-wire can dramatically increase the impedance at W-band due to its parasitic inductance. This impedance could be lowered and matched to 50 Ω by using the circuit section consisting of an open stub and a high impedance line. A two-section matching network was employed to accomplish the broadband matching performance. Figure 6c demonstrated that the designed compensation circuit could recover the performance providing broadband impedance match and lower loss at the W-band. The mismatch increased at higher frequencies due to the increased impedance of the bond-wire. The simulated return loss was higher than 9.1 dB across the full W-band.

3. Experimental Results

3.1. W-Band Waveguide Transition

The designed W-band waveguide package was fabricated in an aluminum split-block configuration. Figure 7a shows the magnified view of the back-to-back transition including the empty spaces for the LNA IC and bias circuits. The transition substrate was mounted on the channel and its two ends were also fixed in the slit region using epoxy. Figure 7b presents the measurement results using a vector network analyzer (VNA; 8510C by Agilent in Santa Clara, CA, USA) with a W-band frequency extender. The simulation results were also plotted for the comparison. The simulation seemed to underestimate the losses by the microstrip line and waveguide sections. The measured return and insertion losses were better than 13.3 dB and 3.5 dB without any in-band resononaces, respectively, over the entire W-band. The insertion loss was lower than 2.6 dB from 75 GHz to 94 GHz. Note that the insertion loss included the loss of the 8.7 mm-long microstrip line and 41.8 mm-long waveguide section.

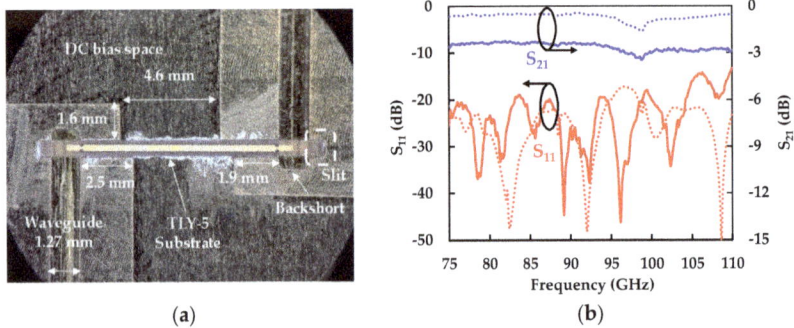

Figure 7. Fabricated W-band transition including the empty spaces for the IC and bias circuits. (**a**) Photograph. (**b**) Measured (solid) and simulated (dotted) S-parameters.

3.2. Bond-Wire Compensation Circuit

In order to verify the effectiveness of the designed bond-wire compensation circuit, the back-to-back transition shown in Figure 7a was slightly modified to have a bond-wire and its compensation circuit as shown in Figure 8a. The microstrip line in the middle would be replaced with the LNA IC later. The Duroid subtrate with the same dielectric constant and thickness to TLY-5 was used in this experiment. The measurement results are presented in Figure 8b, showing a wideband performance with an insertion loss less than 3.9 dB and return loss higher than 10.0 dB from 83.8 GHz to 95.7 GHz.

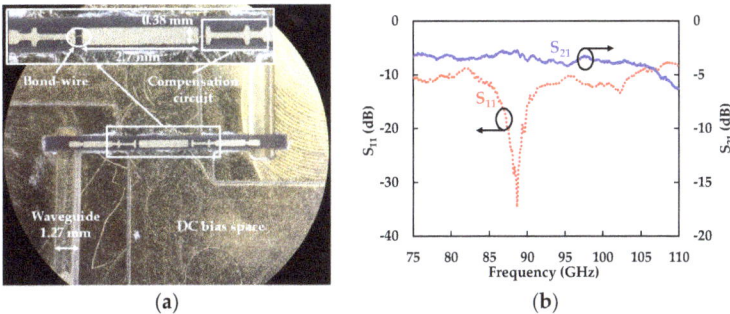

Figure 8. Fabricated W-band transition with bond-wire and its compensation circuit. (**a**) Photograph. (**b**) Measurement results.

3.3. W-Band LNA Module

The fabricated waveguide package was used to build the W-band LNA module as shown in Figure 9. The overall package size was 50 mm × 30 mm × 50 mm. The commercially-available W-band LNA IC (by OMMIC in Paris, France) was mounted and bond-wired to the transition together with the designed bond-wire compensation circuits. Several capacitors were connected in shunt with gate and drain bias lines to suppress low frequency oscillations.

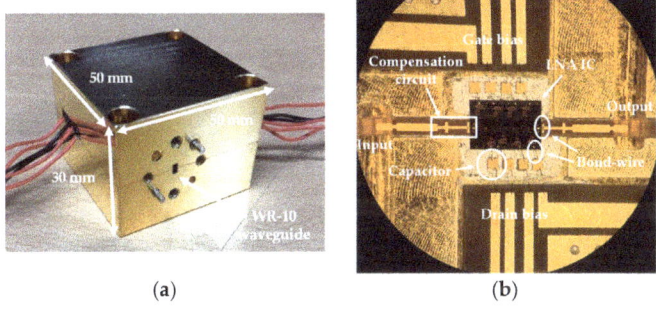

Figure 9. Fabricated W-band LNA module: (**a**) External view and (**b**) magnified view around the LNA IC.

Figure 10a shows the S-parameters of the W-band LNA module measured by the VNA. It achieved the gain ($10\log|S_{21}|^2$) higher than 14.9 dB from 75–105 GHz. It reduced to 12.9 dB at the band edge (110 GHz), which was caused by the LNA and bond-wire compensation circuit. Noise figure was also measured as shown in Figure 10b using a noise figure analyzer (N8975A by Agilent), W-band noise source (NC5110 by Noisecom in Parsippany-Troy Hills, NJ, USA), down-conversion mixer, and spectrum analyzer. The measurement bandwidth was limited to 1.0 GHz from 93.5 GHz to 94.5 GHz by the instruments, where the measured noise figure was less than 4.4 dB. This result appeared to be reasonable, considering the simulated noise figure of the LNA IC (2.8 dB given by the vendor) and the losses of the transitions and bond-wires. Note that the gain measured by the noise figure analyzer was very close to the one by VNA.

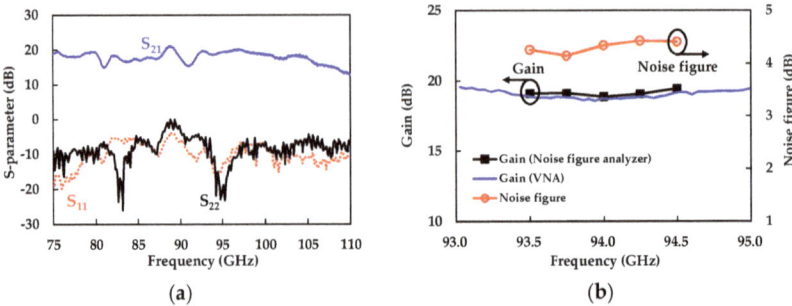

Figure 10. Measured performance of the W-band LNA module. (**a**) S-parameters. (**b**) Noise figure.

4. Conclusions

The broadband waveguide package covering the whole W-band was designed and implemented in this work. The broadband waveguide-to-microstrip transition was presented using an extended E-plane probe transition. The in-band resonances invoked by the cavities, which are formed by the empty spaces for IC and bias circuits, were predicted and removed by designing the asymmetrical structure with open boundary. The bond-wire effect was also simulated and its compensation circuit was designed to recover broadband performance. The fabricated W-band waveguide package exhibited the resonance-free broadband insertion and return losses. The LNA module using the developed waveguide package achieved broadband high gain with a low noise figure at W-band. These results verify that the designed waveguide package can be successfully utilized in the implementation of high performance W-band electronic modules and systems.

Author Contributions: Conceptualization, J.D., W.P. and J.J.; Methodology, W.P. and J.J.; Software, J.D. and W.P.; Validation, J.D. and W.C.; Formal Analysis, J.D.; Data Curation, J.D.; Writing-Original Draft Preparation, J.J.; Writing-Review & Editing, J.D. and J.J.; Visualization, J.D.; Supervision, J.J.; Funding Acquisition, J.J.

Funding: This work was partly supported by the Institute for Information & Communications Technology Planning & Evaluation (IITP) grant, funded by the Korean government (MSIT) (No. 2016-0-00185, Development of ultra-wideband terahertz CW spectroscopic imaging systems based on electronic devices). This work was also partly supported by a grant to the Terahertz Electronic Device Research Laboratory, funded by Defense Acquisition Program Administration and by the Agency for Defense Development (UD180025RD).

Conflicts of Interest: The authors declare no conflict of interest.

References

1. Lee, S.; Joo, J.; Choi, J.; Kim, W.; Kwon, H.; Lee, S.; Kwon, Y.; Jeong, J. W-band multichannel FMCW radar sensor with switching-TX antennas. *IEEE Sens. J.* **2016**, *16*, 5572–5582. [CrossRef]
2. Jeon, Y.; Bang, S. Front-End Module of 18–40 GHz Ultra-Wideband Receiver for Electronic Warfare System. *J. Electromagn. Eng. Sci.* **2018**, *18*, 188–198. [CrossRef]
3. Li, X.; Xiao, J.; Yu, J. Long-Distance Wireless mm-Wave Signal Delivery at W-band. *J. Lightwave Technol.* **2016**, *34*, 661–668. [CrossRef]
4. Lee, S.; Lee, S.; Park, H.; Kim, W.; Kwon, H.; Jeong, J.; Kwon, Y. W-band dual-channel receiver with active power divider and temperature-compensated circuit. *Electr. Lett.* **2016**, *52*, 850–851. [CrossRef]
5. Shahramian, S.; Holyoak, M.J.; Baeyens, Y. A 16-Element W-band Phased-Array Transceiver Chipset With Flip-chip PCB Integrated Antennas for Multi-Gigabit Wireless Data Links. In Proceedings of the IEEE Radio Frequency Integrated Circuits Symposium (RFIC), Phoenix, AZ, USA, 17–19 May 2015; pp. 27–30. [CrossRef]
6. Fujiwara, K.; Kobayashi, T. Low transmission loss, simple and broadband waveguide to microstrip line transducer in *V*-, *E*- and *W*-band. *IEICE Electr. Express* **2017**, *14*, 20170631. [CrossRef]
7. Lee, S.; Son, D.; Kwon, J.; Park, Y. Analysis of a Tapered Rectangular Waveguide for *V* to *W* Millimeter Wavebands. *J. Electromagn. Eng. Sci.* **2018**, *18*, 248–253. [CrossRef]

8. Ma, X.; Xu, R. A broadband *W*-band *E*-plane waveguide to microstrip probe transition. In Proceedings of the IEEE Asia Pacific Microwave Conference, Macau, China, 16–20 December 2008; pp. 1–4. [CrossRef]
9. Li, E.S.; Tong, G.X.; Niu, D.C. Full *W*-band waveguide to microstrip transition with new *E*-plane probe. *IEEE Microw. Wirel. Compon. Lett.* **2013**, *23*, 4–6. [CrossRef]
10. Choe, W.; Kim, J.; Jeong, J. Full *H*-band waveguide-to-coupled microstrip transition using dipole antenna with directors. *IEICE Electr. Express* **2017**, *14*, 1–6. [CrossRef]
11. Kim, J.; Choe, W.; Jeong, J. Submillimeter-Wave Waveguide-to-Microstrip Transitions for Wide Circuits/Wafers. *IEEE Trans. Terahertz Sci. Technol.* **2017**, *7*, 440–445. [CrossRef]
12. Choe, W.; Jeong, J. A Broadband THZ On-Chip Transition Using a Dipole Antenna with Integrated Balun. *Electronics* **2018**, *7*, 236. [CrossRef]
13. Jeong, J.; Kin, D.; Kim, S.; Kwon, Y. *V*-band High-Efficiency Broadband Power Combiner and Power-Combining module Using Double Antipodal Finline Transitions. *Electr. Lett.* **2003**, *39*, 378–379. [CrossRef]
14. Ki, H. A Study on Waveguide to Microstrip Antipodal Transition for 5 G Cellular Systems. *J. Inst. Internet Broadcast. Commun. (JIIBC)* **2015**, *15*. [CrossRef]
15. Zhang, Y.; Shi, S.; Martin, R.D.; Prather, D.W. Broadband SIW to waveguide transition in multilayer LCP substrates at *W* band. *IEEE Microw. Wirel. Compon. Lett.* **2017**, 224–226. [CrossRef]
16. Park, W.; Choe; Lee, K.; Kwon, J.; Jeong, J. Millimeter-wave waveguide transducer using extended *E*-plane probe. *J. Inst. Internet Broadcast. Commun (JIIBC)* **2018**, *18*, 159–165. [CrossRef]
17. Ariffin, A.; Isa, D.; Malekmohammadi, A. Broadband Transition from Microstrip Line to Waveguide Using a Radial Probe and Extended GND Planes for Millimeter-Wave Applications. *PIER Lett.* **2016**, *60*, 95–100. [CrossRef]
18. Aljarosha, A.; Zaman, A.U. A Wideband Contactless and Bondwire-Free MMIC to Waveguide Transition. *IEEE Microw. Wirel. Compon. Lett.* **2017**, *27*, 437–439. [CrossRef]
19. Hassan, E.; Noreland, D.; Wadbro, E.; Berggren, M. Topology Optimisation of Wideband Coaxial-to-Waveguide Transitions. *Sci. Rep.* **2017**, *7*. [CrossRef] [PubMed]
20. Almenti, F.; Mezzanotte, P.; Roselli, L.; Sorrentino, R. An Equivalent Circuit for the Double Bonding Wire Interconnection. In Proceedings of the IEEE MTT-S International Microwave Symposium Digest, Anaheim, CA, USA, 13–19 June 1999; pp. 633–636.
21. Zhou, P.; Zheng, P.; Yu, W.; Sun, H. Design of A *W*-band Low Noise Amplifier Module With MMIC. In Proceedings of the 2012 5th Global Symposium on Millimeter-Waves, Harbin, China, 27–30 May 2012; pp. 5–8. [CrossRef]
22. Hua, Z.; Hongfei, Y.; Peng, D.; Yongbo, S.; Xiaoxi, N.; Zhi, J.; Xinyu, L. A full *W*-band low noise amplifier module for millimeter-wave applications. *J. Semicond.* **2015**, *36*, 095001. [CrossRef]

© 2019 by the authors. Licensee MDPI, Basel, Switzerland. This article is an open access article distributed under the terms and conditions of the Creative Commons Attribution (CC BY) license (http://creativecommons.org/licenses/by/4.0/).

Article

Permittivity of Undoped Silicon in the Millimeter Wave Range

Xiaofan Yang [1], Xiaoming Liu [2,3,*], Shuo Yu [2], Lu Gan [2,3], Jun Zhou [4,*] and Yonghu Zeng [1]

1. State Key Laboratory of Complex Electromagnetic Environment Effects on Electronics and Information System, Luoyang 471003, China
2. School of Physics and Electronic Information, Anhui Normal University, Wuhu 241002, China
3. Anhui Provincial Engineering Laboratory on Information Fusion and Control of Intelligent Robot, Wuhu 241002, China
4. Terahertz Research Centre, School of Electronic Science and Engineering, University of Electronic Science and Technology of China, Chengdu 610054, China
* Correspondence: xiaoming.liu@ahnu.edu.cn (X.L.); zhoujun123@uestc.edu.cn (J.Z.)

Received: 18 July 2019; Accepted: 8 August 2019; Published: 10 August 2019

Abstract: With the rapid development of millimeter wave technology, it is a fundamental requirement to understand the permittivity of materials in this frequency range. This paper describes the dielectric measurement of undoped silicon in the E-band (60–90 GHz) using a free-space quasi-optical system. This system is capable of creating local plane wave, which is desirable for dielectric measurement in the millimeter wave range. Details of the design and performance of the quasi-optical system are presented. The principle of dielectric measurement and retrieval process are described incorporating the theories of wave propagation and scattering parameters. Measured results of a sheet of undoped silicon are in agreement with the published results in the literature, within a discrepancy of 1%. It is also observed that silicon has a small temperature coefficient for permittivity. This work is helpful for understanding the dielectric property of silicon in the millimeter wave range. The method is applicable to other electronic materials as well as liquid samples.

Keywords: millimeter wave; quasi-optics; free space method; undoped silicon; permittivity; temperature

1. Introduction

Dielectric permittivity plays a fundamental role in describing the interaction of electromagnetic waves with matter. The boundary conditions, wave propagation in matter, and wave reflection/transmission on interfaces all involve this parameter. In practical applications, the processes of link budget (wave attenuation), channel characterization (wave dispersion), and multi-path effect (reflection/transmission) are representative examples of wave interaction with media [1]. Understanding these phenomena is a fundamental requirement in communication system design.

With the rising of millimeter wave technology, for example, 5G communication [2], millimeter wave radar and sensing [3], accurate characterization of electronic materials for these applications is of fundamental significance. Although the dielectric properties have been widely investigated in the microwave range or below, the work in the millimeter wave range is much less thorough. The reason is that it has been a long period that the operation frequency of communication systems has been limited to the low frequency range [4]. However, problems such as frequency shift and increased insertion loss are often observed in the millimeter wave circuit design due to inaccurate permittivity [5]. In view of these facts, it is necessary to investigate the dielectric property of materials in the millimeter wave range. Particularly, we restricted our study in the E-band (60–90 GHz), mainly due to the fact that many applications fall in the range, such as the planned unlicensed 64–71 GHz band for 5G communication in the USA [6] and 77 GHz vehicle radar [7]. In addition, we selected silicon as the representative materials for measurement since it is one of the most important semiconductor materials in electronic systems.

There are many publications focused on the complex permittivity of silicon [8–17]. Reference [8] used a capacitive method, which was suitable for measurement in low frequency range. Reference [9] investigated the n-type silicon at 107.3 GHz using a non-focusing free-space system. Afsar and his colleagues [10,11] measured several semiconductor samples using dispersive Fourier transform spectroscopy (DFTS) and presented data over 100–450 GHz [10,11]. It was found that the permittivity slightly increased with frequency from 100 to 180 GHz and then decreased with frequency. In contrast, the loss tangent decreased with the frequency monotonically. However, for silicon of resistivity 11,000 $\Omega\cdot$cm, the loss tangent increases with frequency (see Figure 3 in Reference [11]). The Fabry–Perot resonator was also utilized for silicon characterization over the frequency range of 30–300 GHz [12]. Additionally, it was found that the permittivity had a slight tendency of linear decreasing over the frequency range. An extensive investigation was conducted by Krupka and his colleagues [12–16]. These measurements used resonator techniques at various frequency ranges below 50 GHz. It was demonstrated that the loss tangent decreased with increasing frequency, and in most cases on the order of 10^{-4}. Temperature effects were also systematically investigated over the range of 10–370 K. The real part showed very stable properties over the investigated frequency and temperature ranges, in the worst case in the range of 11.46–11.71. These findings provide valuable reference to the permittivity of silicon.

In these studies the E-band is covered in [12], where it was stated that the refractive index decreased linearly with frequency. Although it is in a small magnitude, such a statement is slightly different from other publications. Considering that many applications have been developed in the E-band, it is best to conduct the measurement in this band. In consideration of these facts, this paper introduces a broadband method for dielectric measurement on semiconductor materials, with measurement conducted on undoped silicon. A detailed description on the condition of creating quasi-plane wave is given. Plane wave is very desirable in free space measurement. The retrieval of the permittivity incorporates the theories of wave propagation and scattering parameters. Temperature dependent characteristics are examined at 20, 25, and 30 °C.

The remaining parts are organized as follows: Section 2 is devoted to a general description on the permittivity and measurement techniques; Section 3 describes the method generating a quasi-plane wave; Section 4 is the retrieval method and Section 5 is the measurement results; the last part Section 6 concludes this work.

2. Permittivity and Measurement Techniques

Dielectric permittivity is a macroscopic description of many microscopic processes, such as dipole relaxation, lattice vibration, and electronic polarization. These processes take effect at different characteristic frequency ranges, as illustrated in Figure 1.

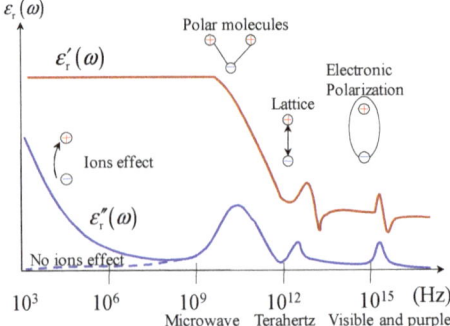

Figure 1. An illustration of frequency dependent dielectric effect.

The permittivity is actually a complex quantity. The imaginary part is due to dielectric loss. One representative mechanism of loss is the dipole relaxation [1]. When a dipole tries to follow the alternating electric field, friction between dipoles would cause loss of energy. Other mechanisms include conductive loss and lattice loss. These mechanisms can be incorporated into a single parameter the complex permittivity

$$\varepsilon_r = \varepsilon'_r - j\varepsilon''_r, \tag{1}$$

where the real part is the relative permittivity and the imaginary part is referred to as the loss factor. Since these mechanisms fall in different frequency ranges, the permittivity exhibits a dependency on frequency

$$\varepsilon_r(\omega) = \varepsilon'_r(\omega) - j\varepsilon''_r(\omega). \tag{2}$$

Referring back to Figure 1, it has to be noted that the ticks on the axes are not to exact scale, only for illustrative purposes.

Methods of dielectric measurement shall vary with frequency due to the frequency dependent and electrical size effects. Several methods can be used for dielectric measurement in the millimeter wave range, such as resonator method, transmission line method, and free space method [18]. The resonator technique is an efficient way for low-loss single frequency measurement. The transmission line method imposes considerable challenge on sample preparation with the increase of frequency. Therefore, the free space system is more preferred for measurement spanning over a whole frequency band.

Free space measurement requires ideally a plane wave system. However, plane wave is merely an ideal model that can be approximated using the far field of an antenna, but the far field would require a space too large to use. Alternatively, we used a quasi-optical (QO) system to create a quasi-plane wave for dielectric measurement. The QO technique is particularly suitable for millimeter wave system due to its low-loss and wideband nature. It is also possible for multi-polarization measurement. In addition, the size of a QO system is controllable.

3. Generating Quasi-Plan Wave

The QO technique was originally developed for radio astronomy for low-loss and wideband receivers [19]. Transferring the QO technique to dielectric measurement is a good attempt. The QO theory is based on Gaussian beam description of an electromagnetic wave. The propagation and refocusing of a Gaussian beam are illustrated in Figure 2. A Gaussian beam can be described using [20]

$$E(r,z) = \frac{w}{w_0} \exp\left(\frac{-r^2}{w^2} - jkz - \frac{j\pi r^2}{\lambda R} + j\phi_0\right), \tag{3}$$

where w_0 is the beam waist located at $z = 0$, w is the beam radius, R is the radius of curvature and ϕ_0 is the phase shift. These parameters can be written as

$$R = z + \frac{1}{z}\left(\frac{\pi w_0^2}{\lambda}\right), \qquad (4)$$

$$w = w_0\left[1 + \left(\frac{\lambda z}{\pi w_0^2}\right)^2\right]^{0.5}, \qquad (5)$$

$$\tan\phi_0 = \frac{\lambda z}{\pi w_0^2}. \qquad (6)$$

It is seen from Equation (4) that the radius of curvature of the wave front is infinite at the beam waist. This is a property of plane wave. However, as the beam propagates, the size of the beam increases with the propagation distance due to the diffraction effect of the electromagnetic wave, gradually forming a spherical wave front. Therefore, a focusing unit has to be used to refocus the beam to a beam waist. The beam transformation by a focusing unit can be analyzed using the ABCD matrix.

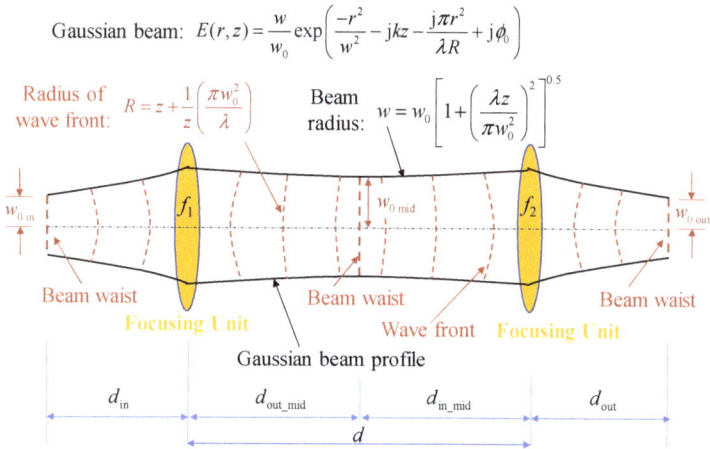

Figure 2. The propagation of a Gaussian beam through free space and the refocusing effect by focusing units.

In a general case, the transform by a focusing unit of focal length f can be calculated by

$$\begin{cases} \frac{d_{out}}{f} = 1 + \frac{d_{in}/f - 1}{(d_{in}/f - 1)^2 + z_c^2/f^2} \\ w_{0\,out} = \frac{w_{0\,in}}{\left[(d_{in}/f - 1)^2 + z_c^2/f^2\right]^{0.5}} \end{cases}, \qquad (7)$$

where d_{in}, d_{out}, $w_{0\,in}$, and $w_{0\,out}$ are the input distance, output distance, input beam waist, and output beam waist, respectively. Additionally, z_c is the confocal distance

$$z_c = \frac{\pi w_{0\,in}^2}{\lambda}. \qquad (8)$$

It is seen that given $d_{in} = f$, the output distance d_{out} is always f, independent of the working frequency. Such a feature enables broadband operation.

For a special case, if two identical focusing units are used and placed $2f$ apart, i.e., $f_1 = f_2 = f$ and $d_1 + d_2 = 2f$ (see Figure 2), the output beam waist will be the same as the input beam waist $w_{0\,out} = w_{0\,in}$. Therefore, the transmitting and receiving horns can be fabricated to be the same. Such a special case gives one a symmetrical structure and is referred to as Gaussian telescope. The system in this paper is in the form of Gaussian telescope.

For a free space system, low cross polarization, low transmission loss, low reflection, and good bandwidth are preferred. To meet these requirements, metallic reflectors can be used as focusing units. Although a dielectric lens can also be used for focusing, the inherent reflection loss and dielectric loss as well as its narrower bandwidth make it less preferred. To obtain low cross polarization, a high-quality horn antenna is always plausible. Corrugated horns are good candidates to the transmitting/receiving units since good bandwidth and polarization purity can be achieved [21]. In addition, the system has to be capable of creating a local plane wave with acceptable transverse size. Additionally, the field should be close enough to a plane wave over a longitudinal distance to ensure that thick samples can also be measured. Such requirements may demand the reflector to be a paraboloidal surface. However, the ellipsoidal surface is more common in a QO system [20].

Considering these requirements, the following parameters were used for the system design (see Table 1). Parameters R_1, R_2, and θ are illustrated in Figure 3a, and f is the equivalent focal length of M1 and M2

$$f = \frac{R_1 R_2}{R_1 + R_2}. \tag{9}$$

Table 1. Key parameters for the Quasi-Optical System, Frequency 90 GHz.

Parameters	w	R_1	R_2	θ	f
Tx-Horn	8.25	-	-	-	-
M1	33.19	500	500	45	250
Sample	32.15	-	-	-	-
M2	33.19	500	500	45	250
Rx-Horn	8.25	-	-	-	-

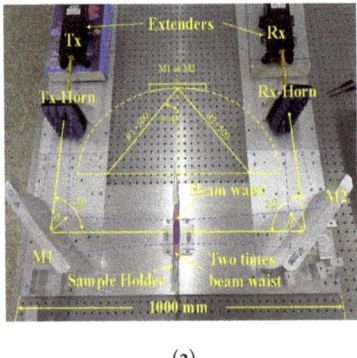 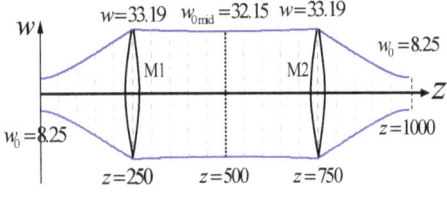

(a) (b)

Figure 3. The photograph of the quasi-optical system for dielectric measurement. (a) The photograph, (b) the beam waist as beam propagation.

The parameter w is the beam radius along propagation, as indicated in Figure 3b.

The fabricated system is shown in Figure 3a, and the beam radius (solid curves) and wave front (dashed curves) are plotted in Figure 3b. The beam radius at the location of M1/M2 is 33.19 mm, which requires the reflector to be 132.76 mm ($4w$ criterion) in length. To accommodate lower frequency of the E-band, the size of the reflector is fabricated to be 200 mm. At the location of the sample, the beam radius is 32.15 mm, which requires the reflector to be 128.60 mm in length. Again, the length of the sample would be suggested to be 200 mm for low frequency consideration.

The dashed lines in Figure 3b are calculated wave fronts using a computer program based on Equation (5). Noticeably, the wave front between M1 and M2 is very close to a plane wave. This is a very good property for dielectric measurement since phase error will be minimized. It is calculated that the phase difference between the edge of the reflector to that of the on-axis phase is only within

10°. The extenders are *VDI VNAX standard*, working at waveguide band WR-12, corresponding to 60–90 GHz. The horns are corrugated ones launching Gaussian beams. The Gaussianities of the horns are obtained by calculating the power coupling between an ideal Gaussian beam and the measured results, giving 98.5%.

4. Retrieval Model

Since we created a quasi-plane wave, the analysis can be conducted using the plane wave method. Such assumption would introduce limited measurement error while significantly reducing the complexity of mathematic manipulation. Reflection on an air-medium interface is an important part in electromagnetism. The reflection model, however, is based on an ideal case where the medium is infinitely long. To apply the ideal model to dielectric measurement, a more precise model has to be built based on the ideal model. This can be simply done by introducing a second interface. The structure can be described by an air-interface-medium-interface-air model, as illustrated in Figure 4.

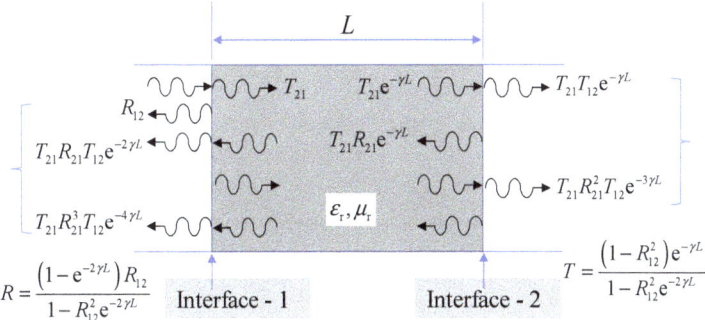

Figure 4. Multi-reflection model.

The interaction of the electromagnetic wave with matter on interfaces 1 and 2 is actually the same as the ideal model. Differently, the reflected wave inside the medium will bounce forward and back. Additionally, these multi-reflection/transmission effects have to be included in calculating the total reflection and transmission coefficients. Using the same mathematical manipulation of [22], the total reflection and transmission coefficients were found to be

$$\begin{cases} R = \frac{(1-e^{-2\gamma L})R_{12}}{1-R_{12}^2 e^{-2\gamma L}} \\ T = \frac{(1-R_{12}^2)e^{-\gamma L}}{1-R_{12}^2 e^{-2\gamma L}} \end{cases}, \quad (10)$$

where the reflection of a plane wave on interface 1 is

$$R_{12} = \frac{1-\sqrt{\varepsilon_r}}{1+\sqrt{\varepsilon_r}}, \quad (11)$$

and the wave propagation coefficient is

$$\gamma = j2\pi \sqrt{\varepsilon_r}/\lambda. \quad (12)$$

Such a structure is identical to a two-port microwave network, where the elements of S_{11}, S_{21} are the reflection and transmission coefficients of port 1, and S_{22}, S_{12} are the reflection and transmission coefficients of port 1, respectively. Interface 1 and interface 2 are similar to port 1 and port 2 of a two-port microwave network, so that the scattering parameters S_{11} and S_{21} measured using a vector network analyzer (VNA) can be correlated to the reflection and transmission coefficients, respectively.

To correlate the transmission coefficient to S_{21}, a measurement on open air has to be conducted prior to sample measurement. In comparison, to correlate the reflection coefficient to S_{11}, a reference measurement of planar metallic plate has to be used. To this end, a link between the wave propagation in a sample and the scattering parameters measured by a vector network analyzer was established.

In Figure 5, we present the first few terms in comparison to the total reflection/transmission coefficients to examine the wave interference at different frequencies. It is seen that the first two or three terms make up most of the reflection/transmission. It is also found that at 160 GHz, the first reflection R1 has a phase of −180°, while the second term R2 and the third term R3 are 0° in phase. Therefore, a destructive interference is formed. At 80 GHz, a constructive interference is formed. Although the third term is out of phase with the first term, the amplitude of R3 is 13 dB smaller than R1, therefore, it would not contribute too much to the interference. The constructive point of the reflection is the destructive point of transmission and vice versa.

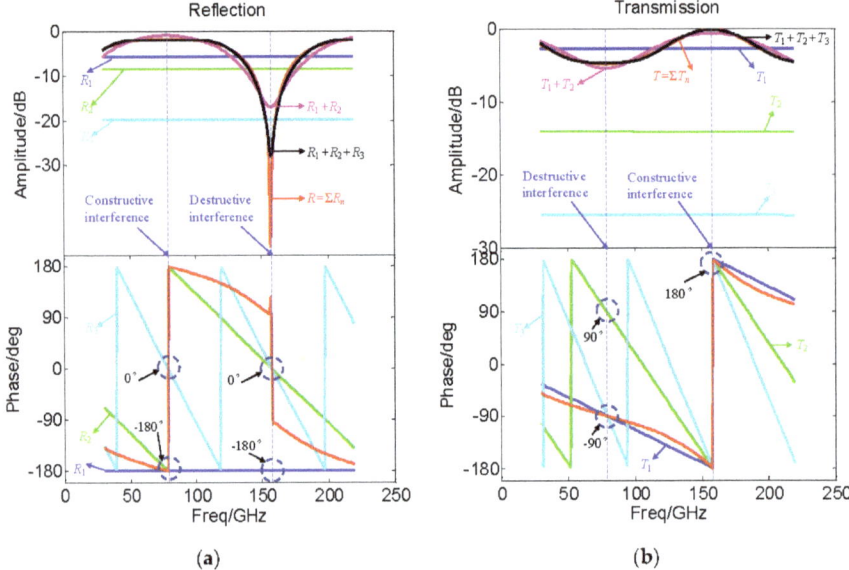

Figure 5. Reflection/transmission of a medium having $\varepsilon_r = 10 - j0.01$. (a) Reflection; (b) transmission.

Retrieve of dielectric permittivity can be fulfilled using Equation (10). For non-magnetic materials, a simple mathematic manipulation can be employed as shown in Reference [23]. The error function can be defined as

$$E(\varepsilon_r) = S_{21} - T. \tag{13}$$

Or using a weighted error function

$$E(\varepsilon_r) = S_{21} - T + w(S_{11} - R). \tag{14}$$

These numerical techniques are available in the literature [22]. In this work, Equation (13) was used as the error function. A flow chart of the Newton–Raphson method for numerical solving is shown in Figure 6.

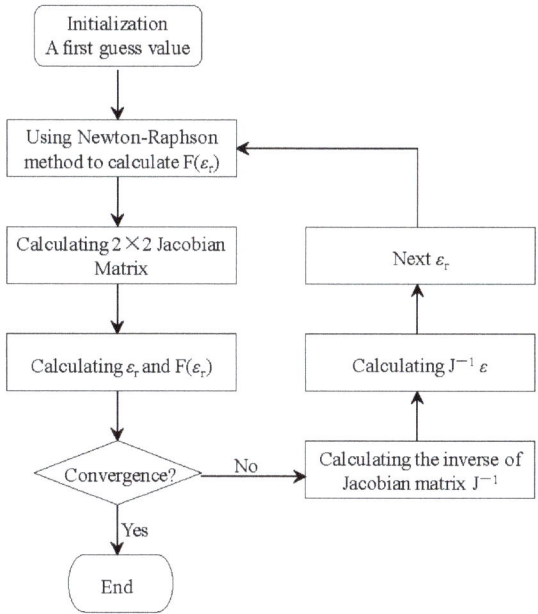

Figure 6. The flow chart of Newton–Raphson method solving for the permittivity.

5. Experiment and Results

The undoped silicon wafer was used for representative measurement. Silicon is a very common material in semiconductor circuits. The provided value of resistivity is 70 kΩ·cm. The diameter of the wafer is 300 mm and the thickness is 660 ± 10 µm. Such a size is sufficiently large for measurement. Both sides of the silicon wafer were polished. Measurement was conducted at 25 °C. The measurement was first conducted on open air. Such a measurement was to remove the background effects, including atmospherical absorption. However, since the atmospherical absorption in the E-band is less than 16 dB/km according to the ITU standard (ITURP.676-11-201609), the total attenuation due to air is less than 0.03 dB given that the path length of this system is smaller than 2 m. In addition, the reference measurement will introduce a phase difference of $2\pi Lc/f$, which should be subtracted from the resulting phase. The corrected raw data of S21 are plotted in Figure 7. It is seen that the transmission response is close to a sinusoidal function. To eliminate the noise, the third and fifth order polynomial fittings were used. Using the Tailor expansion, it is found

$$\begin{cases} T_3(f) \approx a_0 + a_1 f + a_2 f^2 + a_3 f^3 \\ T_5(f) \approx a_0 + a_1 f + a_2 f^2 + a_3 f^3 + a_4 f^4 + a_5 f^5 \end{cases}. \quad (15)$$

If the fifth order polynomial fitting was used, the difference to the third order is only 0.14 dB in the worst case, and the phase difference is only 1°, as shown in Figure 7. The seventh order can also be used, but no significant difference was observed. This is a piece of evidence that the fifth order fitting provides adequate accuracy.

The fitting data were used for permittivity retrieve. In addition, the permittivity versus the frequency is plotted in Figure 8. The real part is plotted in solid line, the imaginary part is plotted in dotted line, and the loss tangent is plotted in dash-dot line. The imaginary part is scaled by 100 and the loss tangent is scaled by 1000.

It is seen that the real part of the permittivity is 11.801 at 60 GHz and gradually decreases to 11.702 at 90 GHz for the third order fitting. For the fifth order fitting, it is 11.755 at 60 GHz and slightly changes to 11.717. The difference of the third order and fifth order fitting is about 0.5%. Such results are in line with the published data of 11.74 [12], and is slightly smaller than the provided value of 11.9 at 1 MHz. The loss tangent of the intrinsic silicon is on the order of 10^{-3}, showing not too much difference between the third and fifth order fitting. However, it has to be mentioned that being constrained by the accuracy of the free space method on the loss factor, it is one order larger than the results in References [14–16]. It is probably due to this reason that the tendency of loss factor with frequency was not exhibited.

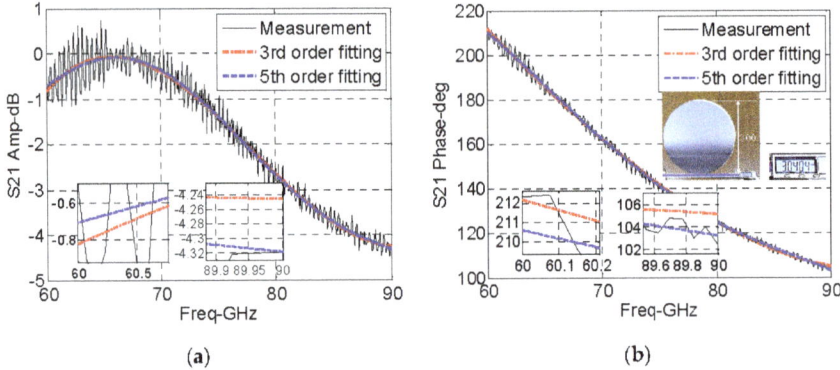

Figure 7. Raw data and fitting results of S21. (a) Amplitude; (b) phase.

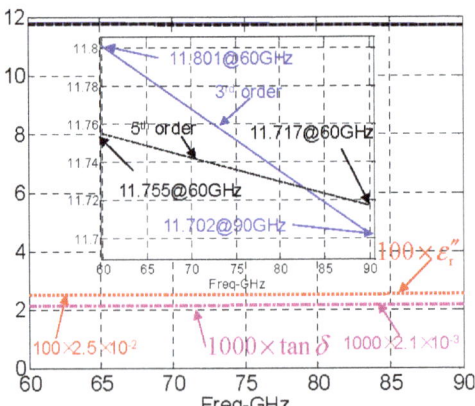

Figure 8. The permittivity in the E-band at 25 °C.

To examine the uncertainty due to the magnitude and phase, we followed the error model in Reference [22], (see Equations (27)–(37) in Section III). The magnitude stability of the *VDI* modules is 0.10 dB, and the phase stability would be 5°. In addition, the fitting model may introduce an extra 0.10 dB magnitude error and 1° phase error. Feeding these values into the error model would give a maximal uncertainty of 5%, and an average error of 3% can be expected.

To investigate the temperature dependent nature of undoped silicon, the temperature was set to 20, 25, and 30 °C. The test bench was placed in a temperature controllable room. The temperature around the sample was measured using a Fiber Optical Sensor (*OMEGA FOB-102*), with the sensitivity of ±0.1 °C. Using the same procedure and taking the fifth order fitting, the retrieved data are plotted in Figure 9. There is a tendency for the real part to decrease with frequency, though not noticeably. It is also found that with the increase in temperature, the real part tends to increase with the increase of temperature. This is in line with the results of References [13,15], where a more broadband temperature range was considered. However, it has to be mentioned that the variation of the permittivity with frequency and temperature is very small, and the overall variation is within 11.75 ± 0.06. It implies that the undoped silicon has a very good temperature property, or very small temperature coefficient. This is a very preferential property for electronic systems.

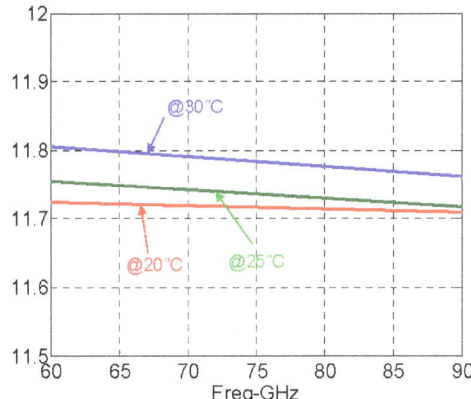

Figure 9. The temperature dependent permittivity in the E-band using the fifth fitting.

A comparison of this work and the published work is shown in Table 2. The extracted data at 25 °C are plotted in comparison in Figure 10. It is noted that as the silicon samples used are from different vendors, differences in resistivity may exist. However, it is seen that the measured results from different groups are in the range of 11.63–11.76, showing a discrepancy of ±0.6%. The measured results in this work are in agreement with the results in the literature [10–14], within a discrepancy of 1.0%. It is also seen that the variation of the permittivity of silicon is very small below 450 GHz. In view of this fact, the dispersion of silicon is very weak, if there is any.

Table 2. A comparison of the measured permittivity of undoped silicon.

Ref.	Freq. (GHz)	Method	Temp.	ε'_r	tanδ	Remarks
[8]	0.03	Capacitive method	77 K	11.7 ± 0.2	-	High-purity silicon
[9]	107.3	Free space non-focusing horn system	23 °C	11–12	< 0.4	n-type silicon
[10]	120–400	Dispersive Fourier transform spectroscopy	27 °C	11.68	$< 2 \times 10^{-3}$	Gen. Diode Silicon
[11]	140	Dispersive Fourier transform spectroscopy	25 °C	11.65–11.71	$< 2 \times 10^{-3}$	-
[12]	30–300	Fabry–Perot resonator	30–300 K	11.74–11.66	$< 2 \times 10^{-4}$	High-purity silicon
[13]	6–18	Cylindrical dielectric resonator	10–370 K	11.46–11.715	$\approx 2 \times 10^{-3}$	High-purity silicon
[14]	1–15	Split post dielectric resonator	-	11.65	1.4×10^{-4}	High-purity silicon
[15]	1–4	Whispering gallery mode resonator	10–350 K	11.46–11.71	$< 10^{-4}$	High purity silicon
[16]	20–50	Whispering gallery mode resonator	Room temperature	11.627–11.653	$< 10^{-4}$	Relative error: ±3%
[17]	9–12	Waveguide measurement	-	11.4–12.0	$< 10^{-3}$	-
This work	60–90	Quasi-optical system	20–30 °C	11.717–11.755	$\approx 2.1 \times 10^{-3}$	Relative error: ±5%

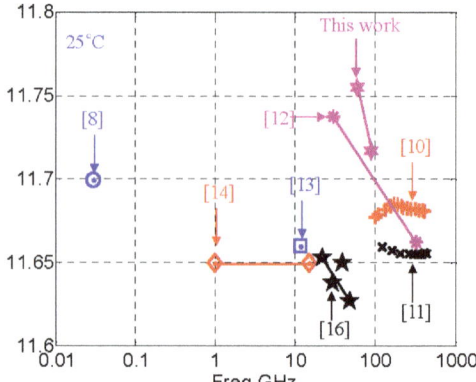

Figure 10. A comparison of the measured permittivity of silicon at different frequency ranges.

6. Conclusions

The permittivity of undoped silicon was measured using a quasi-optical system. The system was designed to create a quasi-plane wave that was preferable for the dielectric measurement millimeter wave range. A multi-layer model was employed for parameter retrieval. The process takes into account multi-path propagation and equals the reflection and transmission coefficients to the scattering parameters in a two-port network. Undoped silicon was measured at the E-band (60–90 GHz). It was found that the undoped silicon exhibited stable permittivity over the whole band and in the temperature range of 20–30 °C. It was found that the measurement was in agreement with the provided value and results in the literature. In addition, it was demonstrated that silicon has very weak dispersion below 450 GHz.

Author Contributions: Data curation, X.Y. and S.Y.; Investigation, X.L.; Methodology, L.G. and J.Z.; Resources, Y.Z.; Writing—original draft, X.L.; Writing—review and editing, X.Y., and J.Z.

Funding: This work was supported by the National Natural Science Foundation of China, Grant Numbers: 61871003, 61601472, 61505022, the Open Project of the State Key Laboratory of Complex Electromagnetic Environment Effects on Electronics and Information System under the number of CEMEE2019Z0203B, and the Natural Science Foundation of Anhui province under the contract number of 1708085QF133.

Conflicts of Interest: The authors declare no conflict of interest.

References

1. Sulyman, A.I.; Nassar, A.T.; Samimi, M.K.; MacCartney, G.R.; Rappaport, T.S.; Alsanie, A. Radio propagation path loss models for 5G cellular networks in the 28 GHZ and 38 GHZ millimeter-wave bands. *IEEE Commun. Mag.* **2014**, *52*, 78–86. [CrossRef]
2. Balanis, C.A. *Advanced Engineering Electromagnetics*, 2nd ed.; Wiley: New York, USA, 2012.
3. Wang, Q.; Liu, S.; Chanussot, J.; Li, X. Scene classification with recurrent attention of VHR remote sensing images. *IEEE Trans. Geosci. Remot.* **2019**, *52*, 1155–1167. [CrossRef]
4. Staecker, P. Microwave industry outlook—Overview. *IEEE Trans. Microw. Theory Tech.* **2002**, *50*, 1034–1036. [CrossRef]
5. Zhang, L.; Zhang, Q.; Hu, C. The influence of dielectric constant on bandwidth of U-notch microstrip patch antenna. In Proceedings of the 2010 IEEE International Conference on Ultra-Wideband, Nanjing, China, 20–23 September 2010.
6. FCC. Use of Spectrum Bands Above 24 GHz For Mobile Radio Services, et al. Report and Order, FCC 16-89. 14 July 2016. Available online: https://www.federalregister.gov/documents/2016/07/14/2016-16620/open-commission-meeting-thursday-july-14-2016 (accessed on 1 August 2019).

7. Kwon, O.-Y.; Cui, C.; Kim, J.-S.; Park, J.-H.; Song, R.; Kim, B.-S. A Compact Integration of a 77 GHz FMCW Radar System Using CMOS Transmitter and Receiver Adopting On-Chip Monopole Feeder. *IEEE Access* **2019**, *7*, 6746–6757. [CrossRef]
8. Dunlap, W.C., Jr.; Watters, R.L. Direct measurement of the dielectric constants of silicon and germanium. *Phys. Rev.* **1953**, *92*, 1396–1397.
9. Kinasewitz, R.T.; Senitzky, B. Investigation of the complex permittivity of n-type silicon at millimeter wavelengths. *J. Appl. Phys.* **1983**, *54*, 3394. [CrossRef]
10. Afsar, M.N.; Button, K.J. Precise millimeter-wave measurements of complex refractive index, complex dielectric permittivity and loss tangent of GaAs, Si, SiO_2, Al_2O_3, BeO, macor, and glass. *IEEE Trans. Microw. Theory Thic.* **1983**, *31*, 217–223. [CrossRef]
11. Afsar, M.N.; Chi, H. Millimeter wave complex refractive index, complex dielectric permittivity and loss tangent of extra high purity and compensated silicon. *Int. J. Infrared Millim. Waves* **1994**, *15*, 1181–1188. [CrossRef]
12. Parshin, V.V.; Heidinger, R.; Andreev, B.A.; Gusev, A.V.; Shmagin, V.B. Silicon as an advanced window material for high power gyrotrons. *Int. J. Infrared Millim. Waves* **1995**, *16*, 863–877. [CrossRef]
13. Krupka, J.; Breeze, J.; Centeno, A.; Alford, N.; Claussen, T.; Jensen, L. Measurements of Permittivity, Dielectric Loss Tangent, and Resistivity of Float-Zone Silicon at Microwave Frequencies. *IEEE Trans. Microw. Theory Thic.* **2006**, *54*, 3995–4001. [CrossRef]
14. Krupka, J.; Kamiński, P.; Kozłowski, R.; Surma, B.; Dierlamm, A.; Kwestarz, M. Dielectric properties of semi-insulating silicon at microwave frequencies. *Appl. Phys. Lett.* **2015**, *107*, 082105. [CrossRef]
15. Krupka, J.; Karcz, W.; Kamiński, P.; Jensen, L. Electrical properties of as-grown and proton-irradiated high purity silicon. *Nucl. Instrum. Methods Phys. Res. Sect. B Beam Interact. Mater. Atoms* **2016**, *380*, 76–83. [CrossRef]
16. Krupka, J.; Kaminski, P.; Jensen, L. High Q-Factor Millimeter-Wave Silicon Resonators. *IEEE Trans. Microw. Theory Tech.* **2016**, *64*, 1–6. [CrossRef]
17. Ismail, K.; Baba, N.H.; Awang, Z.; Esa, M. Microwave Characterization of Silicon Wafer Using Rectangular Dielectric Waveguide. In Proceedings of the 2006 International RF and Microwave Conference, Putra Jaya, Malaysia, 12–14 September 2006.
18. Kaatze, U. Techniques for measuring the microwave dielectric properties of materials. *Metrologia* **2010**, *47*, S91–S113. [CrossRef]
19. Cole, T. IV Quasi-Optical Techniques of Radio Astronomy. *Prog. Opt.* **1977**, *15*, 187–244.
20. Goldsmith, P.F. *Quasioptical Systems: Gaussian Beam Quasioptical Propagation and Applications*; Wiley-IEEE Press: New York, NY, USA, 1998.
21. McKay, J.E.; Robertson, D.A.; Speirs, P.J.; Hunter, R.I.; Wylde, R.J.; Smith, G.M. Compact corrugated feedhorns with high Gaussian coupling efficiency and -60 dB sidelobes. *IEEE Trans. Antennas Propag.* **2016**, *64*, 1. [CrossRef]
22. Baker-Jarvis, J.; Vanzura, E.; Kissick, W. Improved technique for determining complex permittivity with the transmission/reflection method. *IEEE Trans. Microw. Theory Tech.* **1990**, *38*, 1096–1103. [CrossRef]
23. Liu, X.; Chen, H.-J.; Yang, B.; Chen, X.; Parini, C.; Wen, D. Dielectric Property Measurement of Gold Nanoparticle Dispersions in the Millimeter Wave Range. *J. Infrared Millim. Terahertz Waves* **2013**, *34*, 140–151. [CrossRef]

© 2019 by the authors. Licensee MDPI, Basel, Switzerland. This article is an open access article distributed under the terms and conditions of the Creative Commons Attribution (CC BY) license (http://creativecommons.org/licenses/by/4.0/).

Article

Multi-User Linear Equalizer and Precoder Scheme for Hybrid Sub-Connected Wideband Systems

Daniel Castanheira *, Sara Teodoro, Ricardo Simões, Adão Silva and Atilio Gameiro

Instituto de Telecomunicações and DETI, University of Aveiro, 3810-193 Aveiro, Portugal; steodoro@av.it.pt (S.T.); ricardossimoes@ua.pt (R.S.); asilva@av.it.pt (A.S.); amg@ua.pt (A.G.)
* Correspondence: dcastanheira@av.it.pt

Received: 14 March 2019; Accepted: 11 April 2019; Published: 16 April 2019

Abstract: Millimeter waves and massive multiple-input multiple output (MIMO) are two promising key technologies to achieve the high demands of data rate for the future mobile communication generation. Due to hardware limitations, these systems employ hybrid analog–digital architectures. Nonetheless, most of the works developed for hybrid architectures focus on narrowband channels, and it is expected that millimeter waves be wideband. Moreover, it is more feasible to have a sub-connected architecture than a fully connected one, due to the hardware constraints. Therefore, the aim of this paper is to design a sub-connected hybrid analog–digital multi-user linear equalizer combined with an analog precoder to efficiently remove the multi-user interference. We consider low complexity user terminals employing pure analog precoders, computed with the knowledge of a quantized version of the average angles of departure of each cluster. At the base station, the hybrid multi-user linear equalizer is optimized by using the bit-error-rate (BER) as a metric over all the subcarriers. The analog domain hardware constraints, together with the assumption of a flat analog equalizer over the subcarriers, considerably increase the complexity of the corresponding optimization problem. To simplify the problem at hand, the merit function is first upper bounded, and by leveraging the specific properties of the resulting problem, we show that the analog equalizer may be computed iteratively over the radio frequency (RF) chains by assigning the users in an interleaved fashion to the RF chains. The proposed hybrid sub-connected scheme is compared with a fully connected counterpart.

Keywords: massive MIMO; millimeter wave communications; analog precoder; hybrid analog–digital multi-user equalizer; sub-connected architectures

---

1. Introduction

Mobile data traffic has increased over the years and the next generation (5G) is a response to this demand for higher data rates [1,2]. It is expected that 5G will achieve a minimum data rate of 1 Gb/s, 5 Gb/s for high mobility users, and 50 Gb/s for pedestrian users [3,4]. The frequency spectrum used by the mobile communications is currently saturated, so it is necessary to find another spectrum band for mobile applications. In this context the millimeter wave (mmW) band, with its huge available bandwidth [5], with wavelengths from 1 to 10 millimeters [6], can be an interesting solution. Other key technology for future generation communications is massive multiple-input-multiple-output (mMIMO), allowing achievement of higher data rates and better energy efficiency (EE) when compared to the previous generation, 4G-Long Term Evolution (LTE) [7].

The combination of mmW with mMIMO systems enables the use of a large antenna array at the base station (BS) and user terminals (UT) [8]. Massive MIMO, also known as large-scale antenna systems (LSAS), beyond improving EE, can also improve the spectrum efficiency (SE) of the mobile communications systems [9]. SE is independent of the number of antennas employed at the BS and grows with the increase of the number of radio frequency (RF) chains, as discussed in [10]. It is well known that when we have a large number of antennas, it is not feasible to have one dedicated RF chain per antenna, and consequently, a full digital beamforming (BF) architecture is not realistic due to the higher costs and power consumption [9,11]. On the other hand, a system that works only in the analog domain, by employing full analog BF, is not feasible due to the availability of only quantized phase shifters and the constraints on the amplitudes of these analog phase shifters. As a result, the fully analog architecture is normally limited to a single-stream transmission [12]. One possible solution to overcome these limitations is to consider hybrid digital and analog BF where the signal is processed at both analog and digital levels. When this hybrid architecture is compared with the fully digital one, the performance of the hybrid solution is limited by the number of RF chains, but it is possible to design efficient signal processing schemes to achieve a performance close to the fully digital counterpart [13].

1.1. Previous Work on Fully Connected Architectures

Some fully connected hybrid beamforming architectures for narrowband single-user systems were discussed in [14–16]. The work presented in [14] considered a transmit precoding and a receiver combining scheme for mmW mMIMO. In this work the spatial structure of the mmW channels was exploited to design the precoding/combining schemes as a sparse reconstruction problem. In [15] an iterative turbo-like algorithm was proposed that found the near-optimal pair of analog precoder/combiner. A matrix decomposition method that could convert any existing precoder/combining design for the full digital scheme into an analog–digital precoder/combining for the hybrid architecture was addressed in [16]. Approaches for narrowband multi-user systems were also considered in [17–22]. A limited feedback hybrid analog–digital precoding/combining scheme for multi-user systems was addressed in [17]. A heuristic hybrid BF was addressed in [18], where the proposed design could achieve a performance close to the fully digital BF with low-resolution phase shifters. A hybrid analog–digital precoding/combining multi-user system based on the mean-squared error (MSE) was proposed in [19]. The authors of [20] designed an iterative hybrid analog–digital equalizer that efficiently removed the multi-user interferences. In [21], an iterative precoder and combiner design was proposed by exploiting the duality of the uplink and downlink multi-user MIMO channels. In [22], a hybrid beamforming system based on a dual polarized array antenna was proposed for single user systems. The hybrid architecture for either single or multi-user the mmW mMIMO wideband systems was considered in [23–25]. Precoding solutions with codebook design for limited feedback spatial multiplexing in single user wideband mmW were developed in [23]. For the multi-user case, statistical MIMO orthogonal frequency division multiplexing (OFDM) beamformers without instantaneous channel information were designed in [24], where the beams were formed using the dominant eigenvectors to select the main directions. In [25], a downlink MIMO–OFDM hybrid multi-user precoder based on the vector quantization concept was proposed, where the total transmit power was minimized.

1.2. Previous Work on Sub-Connected Architectures

The previous works mainly focused on fully connected architectures. However, sub-connected architectures, where each RF chain is only connected to a subset of the available antennas, is more suited for practical applications due to its lower complexity. Narrowband fixed sub-connected hybrid architectures for single user systems were addressed in [26,27]. The authors of [26] proposed a two-layer optimization method jointly exploiting the interference alignment and fractional programming principles. First, the analog precoder and combiner were optimized via the alternating-direction optimization method and then the precoder and combiner were optimized based on an effective MIMO channel coefficient. In [27] two analog precoder schemes for high and low signal-to-noise ratio (SNR) condition were developed. For multi-user sub-connected narrowband architecture, some approaches were also proposed in [28–32]. In [28], the total achievable rate optimization problem with nonconvex constraints was decomposed into a series of sub-rate optimization problems for each sub-antenna array, and then a successive interference cancelation (SIC) based hybrid precoder was proposed. A low-complexity hybrid precoding and combining design was discussed in [29], where a virtual path was performed to maximize the channel gain and then, based on the effective channel, a zero-forcing precoding was applied to manage the interference. The scheme proposed in [30] efficiently controlled the multi-user interference by sequentially computing the analog part of the equalizer over the RF chains, using a dictionary obtained from the array response vectors. In [31], the Gram–Schmidt (GS) based antenna selection (AS) algorithm was used to obtain an appropriate antenna subset for the overlapped, interlaced, and dynamic architectures. Solutions for wideband sub-connected hybrid architectures were also considered in [32,33]. In [32], solutions for fully connected, fixed, and dynamic subconnected OFDM single user hybrid precoding were designed to maximize the sum rate. Precoding techniques for multi-user downlink massive mmWave MIMO–OFDM systems were proposed in [33]. A unified heuristic design for both fully connected and sub-connected hybrid structures was developed by maximizing the overall spectral efficiency.

1.3. Main Contributions

The previous works considered mainly a sub-connected hybrid architecture for single and multi-user narrowband systems. The works for wideband sub-connected hybrid architectures are very scarce in the literature and they mainly focus on solutions for the downlink considering OFDM modulations. To the best of our knowledge, sub-connected hybrid approaches for uplink of multi-user wideband mmWave massive MIMO systems have yet to be addressed in the literature. Therefore, in this paper we aim to fill this gap and design an efficient hybrid multi-user equalizer combined with a pure analog precoder for sub-connected uplink mmW massive MIMO single-carrier frequency division multiple access (SC–FDMA) systems. We consider single RF UTs employing a low complexity, yet efficient, analog precoder approach based on the knowledge of partial channel state information (CSI), i.e., only a quantized version of the average angle of departure (AoD) of each cluster is considered. The hybrid multi-user linear equalizer employed at the BS is optimized by using the bit-error-rate (BER) as a metric over all the subcarriers. We assume that the digital part of the equalizer is computed on a per subcarrier basis while the analog part is constant over the subcarriers. The analog domain hardware constraints considerably increase the complexity of the corresponding optimization problem. To simplify it, the merit function is first upper bounded, and by leveraging the specific properties of the resulting problem, we show that the analog equalizer may be computed iteratively over the RF chains by assigning the users in an interleaved fashion to the RF chains, using a dictionary built from the array response vectors. The results show that the performance penalty of the sub-connected multi-user equalizer approach to the fully connected counterpart decreases as the number of RF chains increases.

1.4. Organization and Notation

This paper is organized as follows: Section 2 describes the transmitter, channel, and receiver system model. In Section 3 the analog precoder employed at each UT is described, while in Section 4 the sub-connected hybrid analog–digital multi-user equalizer is derived. In Section 5, the main performance results are presented. Finally, the conclusions are discussed in Section 6.

The following notation is used in this paper: boldface uppercase letters, boldface lowercase letters, and italic letters denote matrices, vectors and scalars, respectively. The operations $(.)^T$, $(.)^H$, $(.)^*$, and $tr(.)$ represent the transpose, the Hermitian, the conjugate, and the trace of a matrix, respectively. The operator $\text{diag}(\mathbf{A})$ corresponds to the diagonal entries of the matrix $\mathbf{A}$. The identity matrix of size $N \times N$ is denoted $\mathbf{I}_N$. $\mathbb{E}[.]$ and $\{a_l\}_{l=1}^L$ represent the expectation operator and an L length sequence, respectively. $|a|$ denotes the absolute value of a. $[\mathbf{A}]_{n,l}$ represents the entry of the nth row and lth column of the matrix A. The indices, t, k, and u represent the time domain, subcarrier in the frequency domain, and user terminal, respectively.

2. System Model

In this section, we describe the transmitter, the channel model, and the receiver for the considered uplink massive MIMO mmW SC–FDMA system.

2.1. Transmitter Description

We assume U UTs sharing the same radio resources, each equipped with N_{tx} transmit antennas and with a single RF chain. Figure 1 presents the general schematic of the uth user terminal. Firstly, the time domain N_c-length sequence $\{s_{u,t}\}_{t=0}^{N_c-1}$, with $\mathbb{E}[|s_{u,t}|^2] = 1$, is divided into R data blocks of size $N_s = N_c/R$, where $\{s_{u,t}\}_{t=(r-1)N_s}^{rN_s-1}$ represents the rth data block. Then, this time domain sequence is moved to the frequency domain and the resulting sequence denominated by $\{c_{u,k}\}_{k=(r-1)N_s}^{rN_s-1}$, where $\{c_{u,k}\}_{k=(r-1)N_s}^{rN_s-1}$ is the discrete Fourier transform (DFT) of the time domain sequence $s_{u,t}$. After that, the frequency domain data is interleaved and mapped to the OFDM symbol. To simplify the formulation, we assume that $N_s = N_c$, which means that only a single N_c-length block is considered and assume the identity mapping. Therefore, the frequency domain sequence $\{c_{u,k}\}_{k=0}^{N_c-1}$ is the DFT of the full-time sequence $\{s_{u,t}\}_{t=0}^{N_c-1}$. After the cyclic prefix (CP), an analog precoder $\mathbf{f}_{a,u} \in \mathbb{C}^{N_{tx}}$ is employed. Due to the hardware constraints, we only consider analog phase shifters that force all coefficients of the precoder to have equal magnitude, i.e., $|\mathbf{f}_{a,u}|^2 = 1/N_{tx}$, and furthermore, it is assumed that they are constant over the subcarriers. Therefore, the discrete transmit complex baseband signal $\mathbf{x}_{u,k} \in \mathbb{C}^{N_{tx}}$ of the uth user at subcarrier k can be represented as

$$\mathbf{x}_{u,k} = \mathbf{f}_{a,u} c_{u,k}, \tag{1}$$

where $c_{u,k} \in \mathbb{C}$. The design of the analog precoder coefficients will be presented in the next Section. We assume that the number of users is lower than the number of RF chains (N_{RF}) at the receiver, $U \leq N_{RF}$.

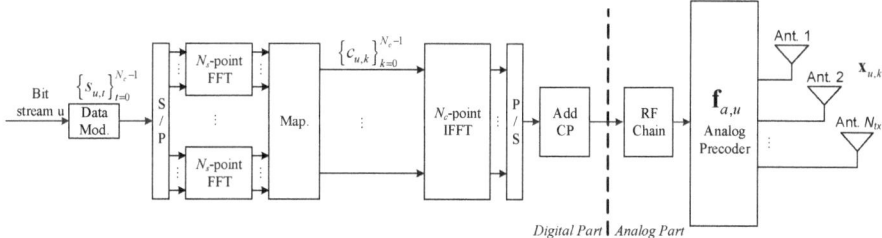

Figure 1. Schematic of the uth user terminal transmitter. CP is cycle prefix, RF is radio frequency, FFT and IFFT are the fast Fourier transform and inverse fast Fourier transform.

2.2. Channel Model Description

We assume a channel given by the sum of the contribution of N_{cl} clusters; each one contributes with N_{ray} propagation paths. The considered delay-d MIMO channel matrix of the uth user can be written as

$$\mathbf{H}_{u,d} = \sqrt{\frac{N_{tx}N_{rx}}{\rho_{PL}}} \sum_{q}^{N_{cl}} \sum_{l}^{N_{ray}} \left(\alpha_{q,l}^u p_{rc}\left(dT_S - \tau_q^u - \tau_{q,l}^u\right) \mathbf{a}_{tx,u}\left(\theta_q^u - \vartheta_{q,l}^u\right) \mathbf{a}_{rx,u}^H\left(\phi_q^u - \varphi_{q,l}^u\right) \right), \tag{2}$$

and the corresponding frequency domain channel matrix $\mathbf{H}_{u,k} \in \mathbb{C}^{N_{rx} \times N_{tx}}$ of the uth user at the kth subcarrier is given by

$$\mathbf{H}_{u,k} = \sum_{d=0}^{D-1} \mathbf{H}_{u,d} e^{-j\frac{2\pi k}{N_c}d}, \qquad (3)$$

where N_{rx} represents the number of receive antennas, ρ_{PL} represents the path-loss between the transmitter and the receiver, $\alpha_{q,l}^u$ is the complex path gain of the lth ray in the qth scattering cluster, and a raised-cosine filter is adopted for the pulse shaping function $p_{rc}(.)$ for T_S-spaced signaling, as in [22]. The qth cluster has a time delay τ_q^u, angles of departure θ_q^u, and arrival ϕ_q^u. Each ray l from qth cluster has a relative time delay $\tau_{q,l}^u$, relative angles of departure $\vartheta_{q,l}^u$ and arrival $\varphi_{q,l}^u$. The paths delay is uniformly distributed in $[0, DT_s]$ where D denotes the length of the CP, and the angles follow the random distribution mentioned in [22], such that $\mathbb{E}[\|\mathbf{H}_{u,d}\|_F^2] = N_{rx}N_{tx}$. Finally, the vectors $\mathbf{a}_{rx,u}$ and $\mathbf{a}_{tx,u}$ represent the normalized receive and transmit array response vectors, respectively. For an N-element uniform linear array (ULA), the array response vector can be given by

$$\mathbf{a}_{ULA}(\theta) = \frac{1}{\sqrt{N}}\left[1, e^{jkp\sin(\theta)}, \ldots, e^{j(N-1)kp\sin(\theta)}\right]^T, \qquad (4)$$

where $k = 2\pi/\lambda$, λ is the wavelength, and p is the inter-element spacing. The channel matrix of the uth user can also be expressed as

$$\mathbf{H}_{u,k} = \mathbf{A}_{rx,u}\mathbf{\Delta}_{u,k}\mathbf{A}_{tx,u}^H, \qquad (5)$$

where $\mathbf{\Delta}_{u,k}$ is a diagonal matrix, with entry (q,l) that corresponds to the path gain of the lth ray in the qth scattering cluster. $\mathbf{A}_{tx,u} = [\mathbf{a}_{tx,u}(\theta_1^u - \vartheta_{1,1}^u), \ldots, \mathbf{a}_{tx,u}(\theta_{N_{cl}}^u - \vartheta_{N_{cl},N_{ray}}^u))]$ and $\mathbf{A}_{rx,u} = [\mathbf{a}_{rx,u}(\phi_1^u - \varphi_{1,1}^u), \ldots, \mathbf{a}_{rx,u}(\phi_{N_{cl}}^u - \varphi_{N_{cl},N_{ray}}^u))]$ hold the transmit and receive array response vectors of the uth user, respectively.

2.3. Receiver Description

At the receiver we consider a hybrid analog–digital sub-connected architecture, where each RF chain is connected into a group of $R = N_{rx}/N_{RF}$ antennas, where N_{RF} is the number of RF chains, as represented in Figure 2. We assume that the number of RF chains is lower than the number of receive antennas, $N_{RF} \leq N_{rx}$.

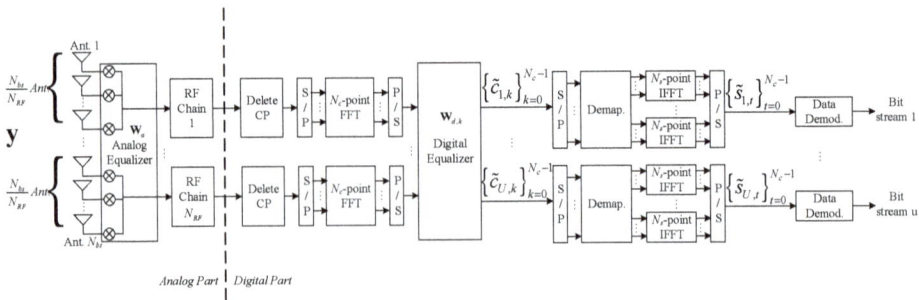

Figure 2. Schematic of the receiver.

The frequency domain received signal at the kth subcarrier $\mathbf{y}_k \in \mathbb{C}^{N_{rx}}$ can be written as

$$\mathbf{y}_k = \sum_{u=1}^{U} \mathbf{H}_{u,k}\mathbf{x}_{u,k} + \mathbf{n}_k = \sum_{u=1}^{U} \mathbf{H}_{u,k}\mathbf{f}_{u,k}c_{u,k} + \mathbf{n}_k, \qquad (6)$$

where $\mathbf{n}_k \in \mathbb{C}^{N_{rx}}$ is the zero mean Gaussian noise with variance σ_n^2, and $\mathbf{x}_{u,k}$ and $\mathbf{f}_{u,k}$ represent the discrete transmit complex baseband signal and analog precoder of the uth user at subcarrier k, respectively. We consider a sub-connected hybrid analog–digital multi-user equalizer to efficiently separate the users, as shown in Figure 2. Initially, the signal is processed through the phase shifters modeled by the vector $\mathbf{w}_{a,r} \in \mathbb{C}^R$, where all elements of $\mathbf{w}_{a,r}$ have equal magnitudes ($|\mathbf{w}_{a,r}(n)|^2 = 1/N_{rx}$). The overall analog matrix $\mathbf{W}_a \in \mathbb{C}^{N_{rx} \times N_{RF}}$ that represents the connection between each subset of N_{rx}/N_{RF} antennas and the corresponding N_{RF} chain, has a block diagonal structure

$$\mathbf{W}_a = \mathrm{diag}\big[\mathbf{w}_{a,1}, \ldots, \mathbf{w}_{a,r}, \ldots, \mathbf{w}_{a,N_{RF}}\big], r = 1, \ldots, N_{RF}. \tag{7}$$

As in the analog precoder, we also assume that the analog part of the equalizer is constant for all subcarriers.

After that, the CP is removed on each RF chain and the signal is moved to the frequency domain by applying the DFT operator. Then, the samples of each subcarrier pass through the digital part of the equalizer modeled by matrix $\mathbf{W}_{d,k} \in \mathbb{C}^{N_{RF} \times U}$. Therefore, the resulting signal at the end of the analog and digital processing equalizer can be written as

$$\tilde{\mathbf{c}}_k = \mathbf{W}_{d,k}^H \mathbf{W}_a^H \mathbf{H}_{eq,k} \mathbf{c}_k + \mathbf{W}_{d,k}^H \mathbf{W}_a^H \mathbf{n}_k, \tag{8}$$

where $\mathbf{H}_{eq,k} = \begin{bmatrix} \mathbf{H}_{1,k}\mathbf{f}_{1,k} & \cdots & \mathbf{H}_{U,k}\mathbf{f}_{U,k} \end{bmatrix} \in \mathbb{C}^{N_{rx} \times U}$ represents the overall equivalent channel between the U users and the receiver, and $\mathbf{c}_k \in \mathbb{C}^U$ denotes the frequency domain transmitted signal of all users at the kth subcarrier.

Finally, the equalized signals are demapped and moved to the time domain by using the inverse DFT, obtaining the estimates $\{\tilde{s}_{u,t}\}_{t=0}^{N_c-1}$ of the uth user transmitted N_c-length data block $\{s_{u,t}\}_{t=0}^{N_c-1}$.

3. Analog Precoder Design

In this section, we design a low complexity analog precoder to be employed at the transmitters. These precoders are computed based on the knowledge of partial CSI, i.e., only a quantized version of the average AoD θ_q^u, $q = 1, \ldots N_{cl}$ of each cluster is used. These angles estimated at the receiver are quantized as

$$\tilde{\theta}_q^u = f_Q(\theta_q^u), q = 1, \ldots N_{cl}, u = 1, \ldots, U, \tag{9}$$

and then sent to the transmitters. In this paper, for the sake of simplicity, we consider uniform quantizers, i.e., the uniform function f_Q has 2^n (n is the number of quantization bits) levels equally spaced between clipping levels $-A_m$ and A_m.

With the knowledge of these quantized angles, user u should start by computing the correlation matrix $\mathbf{R}_u = \widetilde{\mathbf{A}}_{tx,u} \widetilde{\mathbf{A}}_{tx,u}^H$, where the overall matrix $\widetilde{\mathbf{A}}_{tx,u} \in \mathbb{C}^{N_{tx} \times N_{cl}}$ is given as

$$\widetilde{\mathbf{A}}_{tx,u} = [\mathbf{a}_{tx,u}(\tilde{\theta}_1^u), \ldots, \mathbf{a}_{tx,u}(\tilde{\theta}_q^u), \ldots, \mathbf{a}_{tx,u}(\tilde{\theta}_{N_{cl}}^u)], \tag{10}$$

with $\mathbf{a}_{tx,u}(\tilde{\theta}_q^u)$ computed from

$$\mathbf{a}_{tx,u}(\tilde{\theta}_q^u) = \frac{1}{\sqrt{N_{tx}}}\Big[1, e^{jkd\sin(\tilde{\theta}_q^u)}, \ldots, e^{j(N_{tx}-1)kd\sin(\tilde{\theta}_q^u)}\Big]. \tag{11}$$

To compute the analog precoders, we first need to apply the eigenvalue decomposition to the correlation matrix $\mathbf{R}_u$, i.e., $\mathbf{R}_u = \mathbf{U}_{tx,u} \mathbf{\Lambda}_{tx,u} \mathbf{U}_{tx,u}^H$, where $\mathbf{\Lambda}_{tx,u}$ is a diagonal matrix whose elements are the matrix $\mathbf{R}_u$ eigenvalues and $\mathbf{U}_{tx,u}$ a square matrix where the ith column is the ith eigenvector of $\mathbf{R}_u$. Finally, entry n_{tx} of the proposed analog precoders of the uth user is set as

$$\mathbf{f}_{a,u}(n_{tx}) = \frac{1}{\sqrt{N_{tx}}} e^{j\arg(\mathbf{U}_{tx,u}(n_{tx},1))}, n_{tx} = 1, \ldots, N_{tx}, \tag{12}$$

where $\arg(a)$ denotes the argument of complex number a and $\mathbf{U}_{tx,u}(n_{tx}, 1)$, $n_{tx} = 1, \ldots, N_{tx}$ represents entry n_{tx} of the eigenvector corresponding to the largest eigenvalue of the correlation matrix $\mathbf{R}_u$. Hence, the beam follows the best channel direction, improving the transmit/receive link reliability.

4. Multi-User Equalizer Design

In this section, we design a hybrid analog–digital sub-connected equalizer for multi-user mmW mMIMO to be employed at the receiver side. A decoupled transmitter–receiver optimization problem is assumed in this paper, since a joint optimization problem is a very complex task. The overall analog matrix $\mathbf{W}_a$ defined in (7) and the digital matrices $\{\mathbf{W}_{d,k}\}_{k=0}^{N_c-1}$ are optimized by minimizing the BER, which is equivalent to minimize the MSE.

4.1. Problem Formulation

It can be shown that the digital part of the equalizer that minimizes the BER is given by

$$\mathbf{W}_{d,k} = \left(\mathbf{W}_a^H \mathbf{H}_{eq,k} \mathbf{H}_{eq,k}^H \mathbf{W}_a + \sigma_n^2 \mathbf{W}_a^H \mathbf{W}_a\right)^{-1} \mathbf{W}_a^H \mathbf{H}_{eq,k} \tag{13}$$

since it maximizes the overall signal-to-interference-plus-noise-ratio (SINR) of the uth user at time t, $\text{SINR}_{u,t}$, i.e., the SINR relatively to data symbol $s_{u,t}$ [23].

Let us now describe the method to compute the analog part of the considered sub-connected architecture. By using the matrix inversion lemma [34], the overall analog–digital equalizer matrix simplifies to

$$\mathbf{W}_a \mathbf{W}_{d,k} = \mathbf{W}_a \left(\mathbf{W}_a^H \mathbf{W}_a\right)^{-1} \mathbf{W}_a^H \mathbf{H}_{eq,k} \left(\mathbf{H}_{eq,k}^H \mathbf{W}_a \left(\mathbf{W}_a^H \mathbf{W}_a\right)^{-1} \mathbf{W}_a^H \mathbf{H}_{eq,k} + \sigma_n^2 \mathbf{I}\right)^{-1}. \tag{14}$$

Assuming quadrature phase shift keying (QPSK) constellations for simplicity and without loss of generality, the average BER can be written as

$$\text{BER} = \frac{1}{N_c U} \sum_{t=0}^{N_c-1} \sum_{u=1}^{U} Q\left(\sqrt{2\text{SINR}_{u,t}}\right) \tag{15}$$

where Q represents the well-known Q-function and with the $\text{SINR}_{u,t}$ given by

$$\text{SINR}_{u,t}[\mathbf{W}_a] = \left(\frac{\sigma_n^2}{N_c} \sum_{k=0}^{N_c-1} \left[\left(\mathbf{H}_{eq,k}^H \mathbf{W}_a \left(\mathbf{W}_a^H \mathbf{W}_a\right)^{-1} \mathbf{W}_a^H \mathbf{H}_{eq,k} + \sigma_n^2 \mathbf{I}\right)^{-1}\right]_{u,u}\right)^{-1} - 1, \tag{16}$$

where $[\mathbf{A}]_{u,u}$ represents the entry of the uth row and column of the matrix $\mathbf{A}$. From (16) we see that $\text{SINR}_{u,t}$ is independent of the time index. So, we can simplify (15) as

$$\text{BER} = \frac{1}{U} \sum_{u=1}^{U} Q\left(\sqrt{2\text{SINR}_u[\mathbf{W}_a]}\right), \tag{17}$$

with $\text{SINR}_u = \text{SINR}_{u,1} = \text{SINR}_{u,2} = \ldots = \text{SINR}_{u,N_c}$.

The optimization problem to compute the analog part of the equalizer may be mathematically formulated as

$$(\mathbf{W}_a)_{\text{opt}} = \arg \min_{\mathbf{W}_a} \text{BER}[\mathbf{W}_a], \text{ s.t. } \mathbf{W}_a \in \mathcal{W}_a, \tag{18}$$

where $\mathcal{W}_a = \{\mathbf{W}_a : \mathbf{W}_a = \text{diag}[\mathbf{w}_{a,1}, \ldots, \mathbf{w}_{a,r}, \ldots, \mathbf{w}_{a,N_{RF}}], |\mathbf{w}_{a,r}(n)|^2 = 1/N_{rx}\}$ denotes the feasible set for the analog equalizer.

4.2. Proposed Method Derivation

Due to the non-convex nature of the merit function and the constraint imposed in problem (18), it is difficult or even impossible to obtain an analytical solution to the optimization problem at hand. Moreover, as we are considering a multi-user scenario, the resulting average BER is a weighted function of the average BER for each user, making it even harder to obtain a solution to the aforementioned problem. Hence, instead of an exact solution to problem (18) we will derive, in the following, an algorithm to obtain an approximate solution to the previous optimization problem.

Using the exponential upper bound of the Q-function ($Q(x) \leq (1/2)e^{-x^2/2}$), we obtain $Q(x) \leq (1/2)(1+x^2/2)^{-1}$, and as a consequence we have

$$\text{BER} \leq \frac{1}{2U} \sum_{u=1}^{U} \frac{1}{1+\text{SINR}_u}. \tag{19}$$

Replacing Equation (16) in Equation (19), and after some mathematical manipulations, we obtain,

$$\text{BER} \leq \frac{\sigma_n^2}{2UN_c} \sum_{u=1}^{U} \sum_{k=0}^{N_c-1} \left[\mathbf{H}_{eq,k}^H \mathbf{W}_a \left(\mathbf{W}_a^H \mathbf{W}_a \right)^{-1} \mathbf{W}_a^H \mathbf{H}_{eq,k} + \sigma_n^2 \mathbf{I} \right]_{u,u}^{-1}, \tag{20}$$

and then an approximate solution to the optimization problem (18) may be obtained from the following simplified optimization problem

$$\mathbf{W}_a = \arg\min_{\mathbf{W}_a} \sum_{u=1}^{U} \sum_{k=0}^{N_c-1} \left(\mathbf{h}_{eq,u,k}^H \mathbf{W}_a \left(\mathbf{W}_a^H \mathbf{W}_a \right)^{-1} \mathbf{W}_a^H \mathbf{h}_{eq,u,k} + \sigma_n^2 \right)^{-1}, \text{ s.t. } \mathbf{W}_a \in \mathcal{W}_a, \tag{21}$$

where $\mathbf{h}_{eq,u,k} \in \mathbb{C}^{N_{rx}}$ represents the equivalent channel of user u. To solve it, we propose to iteratively compute matrix $\mathbf{W}_a$ column by column, which in practice corresponds to iteratively adding RF chains to the receiver. Let matrix $\mathbf{W}_a^{(i)}$ and vector $\mathbf{w}_a^{(i)}$ denote the first i columns and column i of matrix $\mathbf{W}_a$, respectively; then we can define $\mathbf{W}_a^{(i)} = [\mathbf{W}_a^{(i-1)}, \mathbf{w}_a^{(i)}]$. The aim is to compute $\mathbf{w}_a^{(i)}$ iteratively instead to compute the overall matrix $\mathbf{W}_a$ at once, and thus the optimization problem can be modified as

$$\left(\mathbf{w}_a^{(i)}\right)_{opt} = \arg\min_{\mathbf{w}_a^{(i)}} \sum_{u=1}^{U} \sum_{k=0}^{N_c-1} \left(\mathbf{h}_{eq,u,k}^H \mathbf{W}_a^{(i)} \left(\left(\mathbf{W}_a^{(i)}\right)^H \mathbf{W}_a^{(i)} \right)^{-1} \left(\mathbf{W}_a^{(i)}\right)^H \mathbf{h}_{eq,u,k} + \sigma_n^2 \right)^{-1}, \text{ s.t. } \mathbf{W}_a \in \mathcal{W}_a. \tag{22}$$

From the definition of $\mathbf{W}_a^{(i)}$ and the Gram–Schmidt orthogonalization follows $\mathbf{W}_a^{(i)} (\mathbf{W}_a^{(i)H} \mathbf{W}_a^{(i)})^{-1/2} = [\mathbf{U}^{i-1}, \mathbf{P}^{(i-1)} \mathbf{w}_a^{(i)}]$ [35], where $\mathbf{P}^{(i-1)} = \mathbf{I} - \mathbf{U}^{(i-1)} \mathbf{U}^{(i-1)H}$, and $\mathbf{U}^{(i-1)} = \mathbf{W}_a^{(i-1)} (\mathbf{W}_a^{(i-1)H} \mathbf{W}_a^{(i-1)})^{-1/2}$. Note that $\mathbf{W}_a^{(i)}$ is a block diagonal matrix and therefore $\mathbf{P}^{(i)}$ and $\mathbf{U}^{(i)}$ are also block diagonal. Therefore, the term in the denominator of the merit function of the optimization problem (22) can be simplified to

$$S_{u,k}^{(i)} = \mathbf{h}_{eq,u,k}^H \mathbf{U}^{(i-1)} \mathbf{U}^{(i-1)H} \mathbf{h}_{eq,u,k} + \mathbf{h}_{eq,u,k}^H \mathbf{P}^{(i-1)} \mathbf{w}_a^{(i)} \mathbf{w}_a^{(i)H} \mathbf{P}^{(i-1)H} \mathbf{h}_{eq,u,k} + \sigma_n^2. \tag{23}$$

Notice that the first term in Equation (23) is constant at iteration i, since it does not depend on vector $\mathbf{w}_a^{(i)}$, and is equal to zero for the first iteration. Furthermore, as $\mathbf{W}_a^{(i)} (\mathbf{W}_a^{(i)H} \mathbf{W}_a^{(i)})^{-1} \mathbf{W}_a^{(i)H}$ is a projection matrix it follows that

$$S_{u,k}^{(i)} \leq \mathbf{h}_{eq,u,k}^H \mathbf{h}_{eq,u,k} + \sigma_n^2 \tag{24}$$

with equality if $\mathbf{W}_a^{(i)}(\mathbf{W}_a^{(i)H}\mathbf{W}_a^{(i)})^{-1}\mathbf{W}_a^{(i)H} = \mathbf{I}$. Therefore, if one of the terms of Equation (23) is large, the other one must be small and the following approximation follows,

$$S_{u,k}^{(i)} \approx \max\left\{\begin{array}{l} \mathbf{h}_{eq,u,k}^H \mathbf{U}^{(i-1)} \mathbf{U}^{(i-1)H} \mathbf{h}_{eq,u,k'} \\ \mathbf{h}_{eq,u,k}^H \mathbf{P}^{(i-1)} \mathbf{w}_a^{(i)} \mathbf{w}_a^{(i)H} \mathbf{P}^{(i-1)H} \mathbf{h}_{eq,u,k} \end{array}\right\} + \sigma_n^2. \qquad (25)$$

Nonetheless, due to the independence of the U user channels, if $\mathbf{w}_a^{(i)}$ leads to a large value for the term $\mathbf{h}_{eq,u,k}^H \mathbf{P}^{(i-1)} \mathbf{w}_a^{(i)} \mathbf{w}_a^{(i)H} \mathbf{P}^{(i-1)H} \mathbf{h}_{eq,u,k}$ for user $u_i = u$, then with high probability for the other users $u \neq u_i$, the value of this term will be small. For this reason, $S_{u,k}^{(i)}$ may be approximated as follows

$$S_{u,k}^{(i)} \approx \left\{\begin{array}{ll} \mathbf{h}_{eq,u,k}^H \mathbf{U}^{(i-1)} \mathbf{U}^{(i-1)H} \mathbf{h}_{eq,u,k'} & u \neq u_i \\ \mathbf{h}_{eq,u,k}^H \mathbf{P}^{(i-1)} \mathbf{w}_a^{(i)} \mathbf{w}_a^{(i)H} \mathbf{P}^{(i-1)H} \mathbf{h}_{eq,u,k} + \sigma_n^2, & u = u_i \end{array}\right.. \qquad (26)$$

Therefore, the optimization problem (22) can be approximated by

$$\mathbf{w}_a^{(i)} = \arg\min_{\mathbf{w}_a^{(i)}} \left(\alpha^{(i)} + \sum_{k=0}^{N_c-1}\left(\mathbf{h}_{eq,u_i,k}^H \mathbf{P}^{(i-1)} \mathbf{w}_a^{(i)} \mathbf{w}_a^{(i)H} \mathbf{P}^{(i-1)H} \mathbf{h}_{eq,u_i,k} + \sigma_n^2\right)^{-1}\right), \text{ s.t. } \mathbf{w}_a^{(i)} \in \mathcal{W}_{a,u_i}, \qquad (27)$$

where $\alpha^{(i)} = \sum_{k=0}^{N_c-1} \sum_{u \neq u_i} \left(\mathbf{h}_{eq,u,k}^H \mathbf{U}^{(i-1)} \mathbf{U}^{(i-1)H} \mathbf{h}_{eq,u,k} + \sigma_n^2\right)^{-1}$ is a constant and thus it only depends on the channels $\mathbf{h}_{eq,u,k}$ and matrix $\mathbf{U}^{(i-1)}$ computed in the previous iteration. $\mathcal{W}_{a,u_i}$ represents the column i of the elements of set $\mathcal{W}_a$.

Therefore, the optimization problem (22) may be simplified to

$$\mathbf{w}_a^{(i)} = \arg\min \sum_{k=0}^{N_c-1}\left(\mathbf{h}_{eq,u_i,k}^H \mathbf{P}^{(i-1)} \mathbf{w}_a^{(i)} \mathbf{w}_a^{(i)H} \mathbf{P}^{(i-1)H} \mathbf{h}_{eq,u_i,k} + \sigma_n^2\right)^{-1}, \text{ s.t. } \mathbf{w}_a^{(i)} \in \mathcal{W}_{a,u_i}. \qquad (28)$$

In spite of the previous simplifications, the optimization problem is still non-convex and hard to solve due to the constraint $\mathbf{w}_a^{(i)} \in \mathcal{W}_{a,u_i}$. Therefore, to further simplify it, we replace the set $\mathcal{W}_{a,u_i}$ by the codebook $\mathcal{F}_{a,u_i} = \mathbf{D}_{u_i} \mathbf{A}_{rx,u_i}$, where $\mathbf{D}_{u_i}$ is a block diagonal matrix where all blocks are zero except u_i, which is equal to the identity matrix, i.e., the elements of the codebook are the normalized receiver array response vectors, which leads to the simpler optimization problem

$$\mathbf{w}_a^{(i)} = \arg\min \sum_{k=0}^{N_c-1}\left(\mathbf{w}_a^{(i)H} \mathbf{P}^{(i-1)H} \mathbf{h}_{eq,u_i,k} \mathbf{h}_{eq,u_i,k}^H \mathbf{P}^{(i-1)} \mathbf{w}_a^{(i)} + \sigma_n^2\right)^{-1}, \text{ s.t. } \mathbf{w}_a^{(i)} \in \mathcal{F}_{a,u_i}. \qquad (29)$$

Notice that we need to follow some criterion to associate a given user to iteration i. We propose to do this association using the following mapping between users and iterations $u_i = i \bmod U$, i.e., the U users are interleaved along the iterations. For example, for $N_{RF} = 4$ and $U = 2$, we have $[u_1, u_2, u_3, u_4] = [1, 2, 1, 2]$.

The procedure to obtain the analog part of the equalizer matrix is presented in Algorithm 1. It can be summarized as follows: Firstly we start with user 1, $\mathbf{U}^{(0)} = 0$ (line 1) and compute the projection matrix $\mathbf{P}^{(1)}$ (line 4). After that, we compute the merit function of the optimization problem (29) for each element of the codebook $\mathcal{F}_{a,u_1}$ (lines 5–9). Vector $\mathbf{w}_a^{(1)}$ is set to be equal to the element of codebook $\mathcal{F}_{a,u_1}$ with the lowest value (lines 10–11). Then, the column vector $\mathbf{P}^{(0)}\mathbf{w}_a^{(1)}$ is added to matrix $\mathbf{U}^{(0)}$ to form $\mathbf{U}^{(1)}$ (line 12). With $\mathbf{U}^{(1)}$, the same procedure may be repeated for the other users according to the mapping between users and iterations defined by $u_i = i \bmod U$.

To compute the optimization problem of (29), we need to compute the correlation matrix $\mathbf{w}_a^{(i)H}\mathbf{P}^{(i-1)H}\mathbf{h}_{eq,u_i,k}\mathbf{h}_{eq,u_i,k}^H\mathbf{P}^{(i-1)}\mathbf{w}_a^{(i)}$ for all the elements of the selected codebook $\mathcal{F}_{a,u_1}$, which may be accomplished with the following expression:

$$r_{i,k} = \text{diag}\left(\mathbf{A}_{rx,u_i}^H \mathbf{D}_{u_i} \mathbf{P}^{(i-1)H} \mathbf{h}_{eq,u_i,k} \mathbf{h}_{eq,u_i,k}^H \mathbf{P}^{(i-1)} \mathbf{D}_{u_i} \mathbf{A}_{rx,u_i}\right). \tag{30}$$

Notice that $\mathbf{P}^{(i)}\mathbf{P}^{(i)} = \mathbf{P}^{(i)}$ since $\mathbf{P}^{(i)}$ is an idempotent matrix. Nonetheless, as the Gram–Schmidt procedure may lead to a loss of orthogonality among vectors [35], we use a Gram–Schmidt algorithm with reorthogonalization that amounts to applying two times the projection matrix $\mathbf{P}^{(i)}$ or using $\left(\mathbf{P}^{(i)}\right)^2$ instead of $\mathbf{P}^{(i)}$.

Algorithm 1: The proposed analog–digital multi-user linear equalizer algorithm for sub-connected architecture

---

Inputs: $N_{rx}, N_{RF}, N_c, \mathbf{H}_{eq}$

Analog Part of the equalizer

1: $\mathbf{U}^{(0)} = 0$
2: **for** $i = 1$ to N_{RF} **do**
3: $\quad u_i = i \bmod U$
4: $\quad \mathbf{P}^{(i-1)} = \left(\mathbf{I} - \mathbf{U}^{(i-1)}\mathbf{U}^{(i-1)H}\right)^2$
5: $\quad \mathbf{f}^{(i)} = 0$
6: $\quad$ **for** $k = 1$ to N_c **do**
7: $\quad\quad r_{i,k} = \text{diag}\left(\mathbf{A}_{rx,u_i}^H \mathbf{D}_{u_i} \mathbf{P}^{(i-1)H} \mathbf{h}_{eq,u_i,k} \mathbf{h}_{eq,u_i,k}^H \mathbf{P}^{(i-1)} \mathbf{D}_{u_i} \mathbf{A}_{rx,u_i}\right)$
8: $\quad\quad \mathbf{f}^{(i)} = \mathbf{f}^{(i)} + (r_{i,k} + \sigma_n^2)^{-1}$
9: $\quad$ **end for**
10: $\quad (q, l) = \arg\min \mathbf{f}^{(i)}$
11: $\quad \mathbf{w}_a^{(i)} = \mathbf{P}^{(i-1)}\mathbf{D}_{u_i}\mathbf{a}_{rx,u_i}(\phi_q^{u_i} - \varphi_{q,l}^{u_i})$
12: $\quad \mathbf{U}^{(i)} = \left[\mathbf{U}^{(i-1)}, \mathbf{P}^{(i-1)}\mathbf{w}_a^{(i)}/\|\mathbf{w}_a^{(i)}\|\right]$
13: $\quad \mathbf{W}_a^{(i)} = [\mathbf{W}_a^{(i-1)}, \mathbf{w}_a^{(i)}]$
14: **end for**

Digital part of the equalizer

$\mathbf{W}_{d,k} = \left(\mathbf{W}_a^H \mathbf{H}_{eq,k} \mathbf{H}_{eq,k}^H \mathbf{W}_a + \sigma_n^2 \mathbf{W}_a^H \mathbf{W}_a\right)^{-1} \mathbf{W}_a^H \mathbf{H}_{eq,k}.$

return: $\mathbf{W}_{d,k}, \mathbf{W}_a$

4.3. Complexity Analysis

The steps presented in Algorithm 1 describe the procedure to obtain the analog and digital parts of the proposed equalizer. In the following, the computational complexity of Algorithm 1 is analyzed. Matrix $\mathbf{P}^{(i)}$ at line 4 may be computed with $\mathcal{O}(R^2)$ complexity, with $R = N_{rx}/N_{RF}$, since matrix $\mathbf{U}^{(i-1)}$ is block diagonal and each column has only R non-zero elements. To obtain vector $r_{i,k}$ at line 7, the vector $\mathbf{A}_{rx,u_i}^H \mathbf{D}_{u_i} \mathbf{P}^{(i-1)H} \mathbf{h}_{eq,u_i,k}$ must be calculated. As both $\mathbf{D}_{u_i}$ and $\mathbf{P}^{(i)}$ are block diagonal with N_{RF} blocks of size $R = N_{rx}/N_{RF}$ and $\mathbf{D}_{u_i}$ is a block diagonal matrix where all blocks are zero except u_i, which is equal to the identity matrix, then the product $\mathbf{D}_{u_i}\mathbf{P}^{(i-1)H}\mathbf{h}_{eq,u_i,k}$ requires $\mathcal{O}(R^2)$ complexity. As the resulting vector has only R non-zero elements and $\mathbf{A}_{rx,u_i}$ is a matrix with dimension $N_{rx} \times N_{cl}N_{ray}$, $\mathbf{A}_{rx,u_i}^H \mathbf{D}_{u_i} \mathbf{P}^{(i-1)H} \mathbf{h}_{eq,u_i,k}$ may be computed with $\mathcal{O}(R^2 + N_{cl}N_{ray}R)$ complexity. As the computation done in line 7 is repeated for all N_c subcarriers and all RF chains, the overall complexity of lines 6–9 is $\mathcal{O}(N_c(R^2 + N_{cl}N_{ray}))$. Line 10 requires the computation of the minimum of vector $\mathbf{f}^{(i)} \in \mathbb{R}^{N_{cl}N_{ray}}$, whose complexity is $\mathcal{O}(N_{cl}N_{ray})$. In line 11 we compute vector $\mathbf{w}_a^{(i)}$ with complexity $\mathcal{O}(R^2)$. The complexity of line 12 is identical to line 11. As the previous must be repeated for all RF chains, the overall complexity is N_{RF} times the individual complexities previously described. The computation of the digital equalizer must be performed for all subcarriers and requires the inversion of a matrix with size $N_{RF} \times N_{RF}$, resulting in a complexity $O(N_c N_{RF}^3)$. Therefore, the computational complexity of the proposed algorithm is linear in the number of subcarriers, quadratic with the number of antennas per RF chain, and cubic with the number RF chains. The full connected architecture would require a complexity scaling quadratically with the number of antennas and for the full-digital architecture the complexity is a cubic function of the number of antennas.

5. Performance Results

In this section, we evaluate the performance of the proposed multi-user linear equalizer and precoder scheme designed for hybrid sub-connected wideband mmW systems.

The carrier frequency was set to 72 GHz and for each user, the clustered wideband channel model was considered, as discussed in Section 2, with five clusters $N_{cl} = 5$, all with the same average power, such that $\mathbb{E}[\|\mathbf{H}_{u,d}\|_F^2] = N_{rx} N_{tx}$, and each one contributed with $N_{ray} = 3$ propagation path. The path delays were uniformly distributed in the CP interval. We considered a ULA with antenna element spacing set to half-wavelength, but it should be emphasized that the schemes proposed in this paper can be applied to any antenna arrays. The azimuth angles of departure and arrival had a Laplacian distribution as in [20] and were considered to have an angle spread of 10° for both the transmitter and receiver. It was assumed a QPSK modulation, perfect synchronization, and that the CSI is known at the receiver side. At the transmitter, only a quantized version of the average angle of departure of each cluster was known. We assumed $N_c = 64$ subcarriers, and the CP was set to be a quarter of the number of subcarriers, such that $D = N_c/4 = 16$. We considered a Monte-Carlo simulation with a length of 100,000 SC–FDMA blocks. The average BER was considered the performance metric, presented as a function of E_b/N_0, where E_b is the average bit energy and N_0 is the one-sided noise power spectral density. We considered that the average E_b/N_0 was identical for all the users u and is given by $E_b/N_0 = 1/(2\sigma_n^2)$.

We considered that each transmitter had a single RF chain and was equipped with $N_{tx} = 8$ antennas. At the receiver side, it was assumed that a sub-connected architecture existed, where each RF chain was connected to a group of $R = N_{rx}/N_{RF}$ antennas, with $N_{rx} = 16$ antennas. The results were compared with the fully connected counterpart that could be obtained from the proposed one by relaxing the optimization constraints, since each RF chain was connected to all antennas. The main simulation parameters are presented in Table 1.

Table 1. Main simulation parameters.

Parameter	Values
Number of user (U)	2, 4, and 8
Number of Receive antennas (N_{rx})	16
Number of Transmit antennas (N_{tx})	8
Number of RF chains (R)	2, 8, and 16
Number of subcarriers (N_c)	64
Cyclic prefix	16
Number of quantization bits (n)	2, 4, and 6
Carrier frequency	72 GHz
Number of clusters (N_{cl})	5
Number of rays per cluster (N_{ray})	3

Figure 3 depicts the results for the proposed hybrid sub-connected multi-user equalizer with the analog precoder for two, four, and eight users. In this figure, it was assumed perfect knowledge of the average AoD of each cluster. It was also assumed eight RF chains existed, which meant that each one was connected to two antennas. As it can be seen in Figure 3, the performance of both sub- and fully connected improved as the number of users decreased, as expected, since the multi-user equalizer had to deal with less interference and the available degrees of freedom could be used to provide more diversity. We could also observe a performance penalty of the sub-connected approach against the fully connected one of approximately 2 dB, independently of the number of users, at a target BER of 10^{-3}. This was because the number of connections of the fully-connected architecture was larger than the number of connections for the sub-connected architecture and, as expected, the result for the fully-connected one was better than for the sub-connected architecture. The worst performance, for both fully and sub-connected approaches, was obtained for the full load case, i.e., when the number of users was equal to the number of RF chains $N_{RF} = U$.

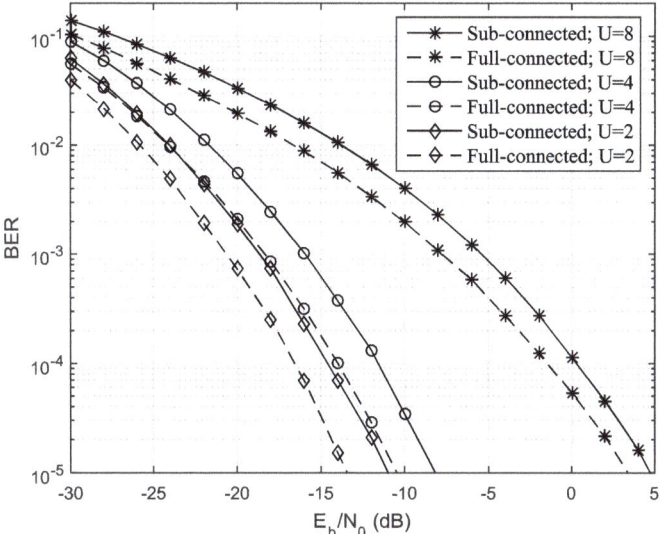

Figure 3. Performance of the proposed hybrid sub-connected scheme for $U \in \{2, 4, 8\}$. BER is bit-error-rate.

In Figure 4 we present results for different numbers of RF chains and two users. If the number of antennas (R) connected to each RF chain was reduced, we verified that the penalty for fully digital approach decreased. We could observe a penalty of approximately 3 dB, 2 dB, and 0 dB for $(R = 8, N_{RF} = 2)$, $(R = 2, N_{RF} = 8)$, and $(R = 1, N_{RF} = 16)$, respectively (BER of 10^{-3}). This happened because more RF chains and consequently less antennas per chain increase the number of available degrees of freedom of the sub-connected architecture. For the extreme case of $R = 1$ (one RF chain per antenna) the curve obtained for the sub-connected approximately overlapped the one obtained for the fully connected.

As seen in Figures 5 and 6, we evaluated the impact of imperfect knowledge of the average AoD at the transmitter side. To compute the analog precoders we assumed the knowledge of only a quantized version of the average AoD of each cluster, as discussed in Section 3. We presented results for $n = [2, 4, 6]$ quantization bits. Figures 5 and 6 depict the results for two and eight users, respectively. As expected, increasing the number of quantization bits improved the performance of the proposed sub-connected scheme and tended to the one achieved for perfect knowledge of the average AoD ($n = \infty$) for both cases $U = 2, 8$. When the number of bits in the quantizer was lower, the performance was worse compared to the perfect curve. In Figure 5 we can observe a performance penalty, for BER of 10^{-3}, of approximately 5 dB, 1.5 dB, and 0 dB, for $n = 2, 4$, and 6, respectively. This means that a very limited number of bits for the quantization of the average AoD of each cluster was enough to get a performance close to the perfect case. Since the mmW channels were usually sparse, the amount of information needed to be fed back from the BS to the UTs was small.

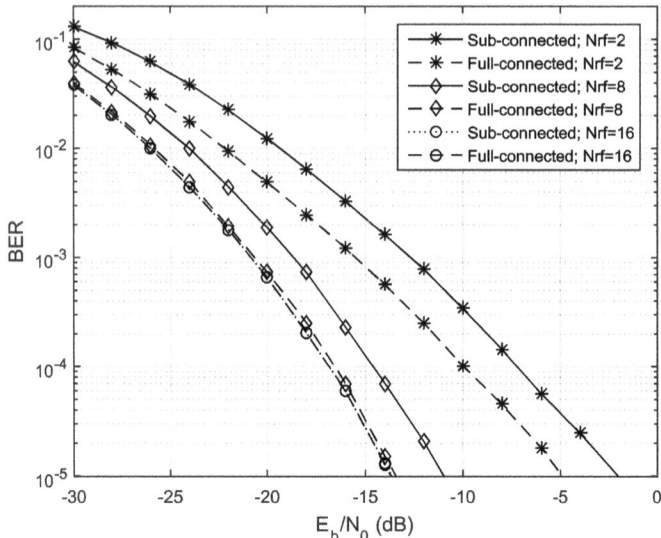

Figure 4. Performance of the proposed hybrid sub-connected schemes for $N_{RF} \in \{2, 8, 16\}$ and $U = 2$.

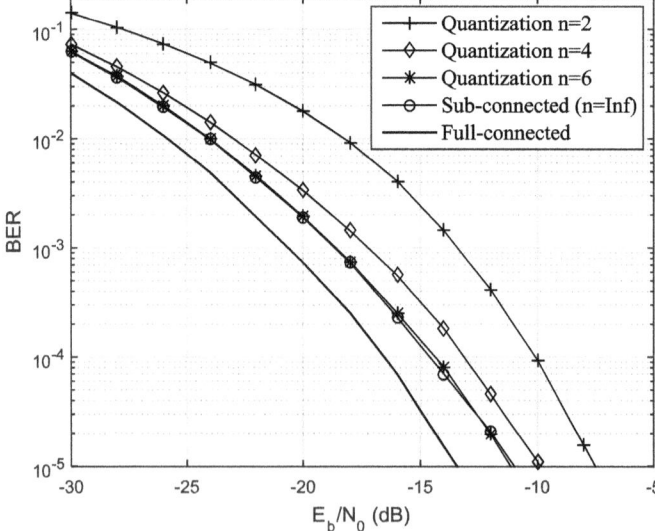

Figure 5. Performance of the proposed hybrid sub-connected scheme for different numbers of quantization bits of the average AoD, $U = 2$.

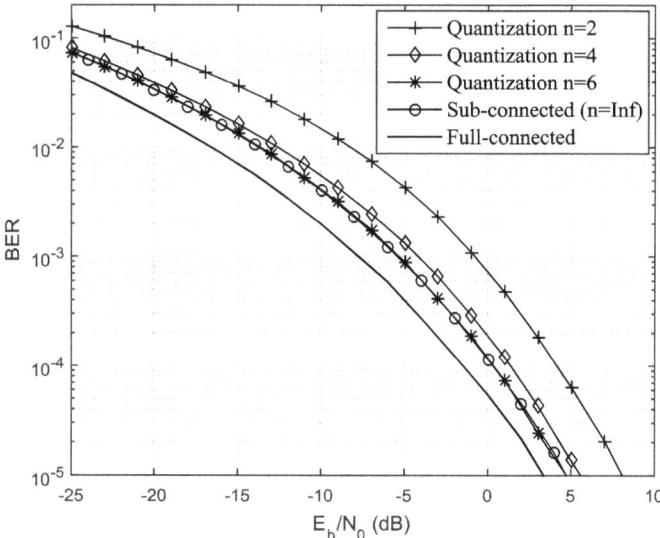

Figure 6. Performance of the proposed hybrid sub-connected scheme for different numbers of quantization bits of the average AoD, $U = 8$.

6. Conclusions

In this paper, we proposed an analog precoder combined with an efficient hybrid analog–digital multi-user equalizer for sub-connected mmW massive MIMO SC–FDMA systems. At the UT, we proposed a low complexity pure analog precoder that requires the knowledge of a quantized version of the average AoD of each cluster. At the BS, a hybrid analog–digital multi-user equalizer was developed for a sub-connected architecture. It was assumed that the analog part was constant over all subcarriers, while the digital part was computed on a per subcarrier basis. We considered a minimum MSE-based equalizer for the digital part, and the analog part was optimized using the average bit-error-rate of all subcarriers as metric. In order to simplify the optimization problem at hand, the merit function was first upper bounded and then, due to the specific properties of the resulting problem, we showed that the analog part of the hybrid equalizer may be computed iteratively over the RF chains by assigning the users in an interleaved fashion to the RF chains.

The numerical results show that the proposed wideband hybrid multi-user linear equalizer is quite efficient at removing the multi-user interference, and the performance tends to the one achieved by the fully connected counterpart as the number of RF chains increases. Furthermore, only a few bits are required for the quantization of the average AoD of each cluster to obtain a performance close to the perfect case. The small performance gap between the proposed sub-connected approach and the fully connected one, together with the lower complexity, make it a very interesting choice for practical systems.

Author Contributions: Conceptualization, D.C. and A.S.; software, S.T. and R.S.; validation, S.T. and R.S.; formal analysis, D.C.; investigation, S.T. and R.S.; writing—original draft preparation, D.C., A.S., and A.G.; writing review and editing, D.C., A.S., and A.G.; funding acquisition, A.S.

Funding: This work is supported by the European Regional Development Fund (FEDER), through the Competitiveness and Internationalization Operational Program (COMPETE 2020) of the Portugal 2020 framework, Regional OP Centro (CENTRO 2020), Regional OP Lisboa (LISBOA 14-20), and by FCT/MEC through national funds, under Project MASSIVE5G (AAC n° 02/SAICT/2017), and in part by the European Structural and Investment Funds (FEEI) through the Competitiveness and Internationalization Operational Program—COMPETE 2020 and by the National Funds through the Foundation for Science and Technology under the Project RETIOT under Grant POCI-01-0145-FEDER-016432.

Conflicts of Interest: The authors declare no conflict of interest.

References

1. Demestichas, P.; Georgakopoulos, A.; Karvounas, D.; Tsagkaris, K.; Stavroulaki, V.; Lu, J.; Xiong, C.; Yao, J. 5G on the Horizon: Key Challenges for the Radio-Access Network. *IEEE Veh. Technol. Mag.* **2013**, *8*, 47–53. [CrossRef]
2. Hossain, E.; Rasti, M.; Tabassum, H.; Abdelnasser, A. Evolution toward 5G multi-tier cellular wireless networks: An interference management perspective. *IEEE Wirel. Commun.* **2014**, *21*, 118–127. [CrossRef]
3. Roh, W.; Seol, J.; Park, J.; Lee, B.; Lee, J.; Kim, Y.; Cho, J.; Cheun, K.; Aryanfar, F. Millimeter-wave beamforming as an enabling technology for 5G cellular communications: Theoretical feasibility and prototype results. *IEEE Commun. Mag.* **2014**, *52*, 106–113. [CrossRef]
4. Bogale, T.E.; Le, L.B. Massive MIMO and mmWave for 5G Wireless HetNet: Potential Benefits and Challenges. *IEEE Veh. Technol. Mag.* **2016**, *11*, 64–75. [CrossRef]
5. Swindlehurts, A.; Ayanoglu, E.; Heydari, P.; Capolino, F. Millimeter-wave massive MIMO: The next wireless revolution? *IEEE Commun. Mag.* **2014**, *52*, 56–62. [CrossRef]
6. Pi, Z.; Khan, F. An introduction to millimeter-wave mobile broadband systems. *IEEE Commun. Mag.* **2011**, *49*, 101–107. [CrossRef]
7. Prasad, K.; Hossain, E.; Bhargava, V.K. Energy efficiency in massive MIMO-based 5G networks: Opportunities and challenges. *IEEE Wirel. Commun.* **2017**, *24*, 86–94. [CrossRef]
8. Vu, T.K.; Liu, C.-F.; Bennis, M.; Debbah, M.; Latva-Aho, M.; Hong, C.S. Ultra-Reliable and Low Latency Communication in mmWave-Enabled Massive MIMO Networks. *IEEE Commun. Lett.* **2017**, *21*, 2041–2044. [CrossRef]
9. Han, S.; Lin, C.; Rowell, C.; Xu, Z.; Wang, S.; Pan, Z. Large scale antenna system with hybrid digital and analog beamforming structure. In Proceedings of the 2014 ICC—2014 IEEE International Conference on Communication Workshop (ICC), Sydney, NSW, Australia, 10–14 June 2014; pp. 842–847.
10. Tan, W.; Matthaiou, M.; Jin, S.; Li, X. Spectral Efficiency of DFT-Based Processing Hybrid Architectures in Massive MIMO. *IEEE Wirel. Commun. Lett.* **2017**, *6*, 586–589. [CrossRef]
11. Roze, A.; Crussiere, M.; Helard, M.; Langlais, C. Comparison between a hybrid digital and analog beamforming system and a fully digital Massive MIMO system with adaptive beamsteering receivers in millimeter-Wave transmissions. In Proceedings of the 2016 International Symposium on Wireless Communication Systems (ISWCS), Poznan, Poland, 20–23 September 2016; pp. 86–91.
12. Alkhateeb, A.; El Ayach, O.; Leus, G.; Heath, R.W. Channel Estimation and Hybrid Precoding for Millimeter Wave Cellular Systems. *IEEE J. Sel. Top. Process.* **2014**, *8*, 831–846. [CrossRef]
13. Alkhateeb, A.; Mo, J.; González-Prelcic, N.; Heath, R., Jr. MIMO precoding and combining solutions for millimeter-wave systems. *IEEE Commun. Mag.* **2014**, *52*, 122–131. [CrossRef]
14. Ayach, O.; Rajagopal, S.; Surra, S.; Pi, Z.; Heath, R. Spatially sparse precoding in millimeter wave MIMO systems. *IEEE Trans. Wirel. Commun.* **2014**, *13*, 1499–1513. [CrossRef]
15. Gao, X.; Dai, L.; Yuen, C.; Wang, Z. Turbo-Like Beamforming Based on Tabu Search Algorithm for Millimeter-Wave Massive MIMO Systems. *IEEE Trans. Veh. Technol.* **2016**, *65*, 5731–5737. [CrossRef]
16. Ni, W.; Dong, X.; Lu, W.S. Near-optimal hybrid processing for massive MIMO systems via matrix decomposition. *IEEE Trans. Signal Process.* **2017**, *65*, 3922–3933. [CrossRef]
17. Alkhateeb, A.; Leus, G.; Heath, R.W. Limited feedback hybrid precoding formulti-user millimeter wave systems. *IEEE Trans. Wirel. Commun.* **2015**, *14*, 6481–6494. [CrossRef]
18. Sohrabi, F.; Yu, W. Hybrid digital and analog beamforming design for large-scale antenna Arrays. *IEEE J. Sel. Top. Signal Process.* **2016**, *10*, 501–513. [CrossRef]
19. Nguyen, D.H.N.; Le, L.B.; Le-Ngoc, T.; Heath, R.W. Hybrid MMSE Precoding and Combining Designs for mmWave Multiuser Systems. *IEEE Access* **2017**, *5*, 19167–19181. [CrossRef]
20. Magueta, R.; Castanheira, D.; Silva, A.; Dinis, R.; Gameiro, A. Hybrid Iterative Space-Time Equalization for Multi-User mmW Massive MIMO Systems. *IEEE Trans. Commun.* **2017**, *65*, 608–620. [CrossRef]
21. Wang, Z.; Li, M.; Tian, X.; Liu, Q. Iterative hybrid precoder and combiner design for mmWave multiuser MIMO systems. *IEEE Commun. Lett.* **2017**, *21*, 1581–1584. [CrossRef]

22. Sung, L.; Park, D.; Cho, D. Limited feedback hybrid beamforming based on dual polarized array antenna. *IEEE Commun. Lett.* **2018**, *22*, 1486–1489. [CrossRef]
23. Alkhateeb, A.; Heath, R.W. Frequency selective hybrid precoding for limited feedback millimeter wave systems. *IEEE Trans. Commun.* **2016**, *64*, 1801–1818. [CrossRef]
24. Lin, Y. Hybrid MIMO-OFDM beamforming for wideband mmWave channels without instantaneous feedback. *IEEE Trans. Signal Process.* **2018**, *66*, 5142–5151. [CrossRef]
25. Kong, L.; Han, S.; Yang, C. Hybrid precoding with rate and coverage constraints for wideband massive MIMO systems. *IEEE Trans. Wirel. Commun.* **2018**, *17*, 4634–4647. [CrossRef]
26. He, S.; Qi, C.; Wu, Y.; Huang, Y. Energy-Efficient Transceiver Design for Hybrid Sub-Array Architecture MIMO Systems. *IEEE Access* **2016**, *4*, 9895–9905. [CrossRef]
27. Li, N.; Wei, Z.; Yang, H.; Zhang, X.; Yang, D. Hybrid precoding for mmWave massive MIMO systems with partially connected structure. *IEEE Access* **2017**, *5*, 15142–15151. [CrossRef]
28. Gao, X.; Dai, L.; Han, S.; Chih-Lin, I.; Heath, R.W. Energy-efficient hybrid analog and digital precoding for mmWave MIMO systems with large antenna arrays. *IEEE J. Sel. Areas Commun.* **2016**, *34*, 998–1009. [CrossRef]
29. Li, A.; Masouros, C. Hybrid Analog-Digital Millimeter-Wave MU-MIMO Transmission with Virtual Path Selection. *IEEE Commun. Lett.* **2017**, *21*, 438–441. [CrossRef]
30. Magueta, R.; Mendes, V.; Castanheira, D.; Silva, A.; Dinis, R.; Gameiro, A. Iterative Multiuser Equalization for Subconnected Hybrid mmWave Massive MIMO Architecture. *Wirel. Commun. Mob. Comput.* **2017**, *2017*, 1–13. [CrossRef]
31. Hu, C.; Zhang, J. Hybrid precoding design for adaptive sub-connected structures in millimeter-wave MIMO systems. *IEEE Syst. J.* **2019**, *13*, 137–146. [CrossRef]
32. Park, S.; Alkhateeb, A.; Heath, R.W. Dynamic subarrays for hybrid precoding in wideband mmWave MIMO systems. *IEEE Trans. Wirel. Commun.* **2017**, *16*, 2907–2920. [CrossRef]
33. Sohrabi, F.; Yu, W. Hybrid analog and digital beamforming for mmWave OFDM large-scale antenna arrays. *IEEE J. Sel. Areas Commun.* **2017**, *35*, 1432–1443. [CrossRef]
34. Palomar, D.; Jiang, Y. *MIMO Transceiver Design via Majorization Theory*; Now: Boston, MA, USA, 2007.
35. Giraud, L.; Langou, J.; Rozložník, M. The loss of orthogonality in the Gram-Schmidt orthogonalization process. *Comput. Math. Appl.* **2005**, *50*, 1069–1075. [CrossRef]

© 2019 by the authors. Licensee MDPI, Basel, Switzerland. This article is an open access article distributed under the terms and conditions of the Creative Commons Attribution (CC BY) license (http://creativecommons.org/licenses/by/4.0/).

Article

Multi-Backup Beams for Instantaneous Link Recovery in mmWave Communications

Adel Aldalbahi

Department of Electrical Engineering, King Faisal University, Hofuf 31982, Saudi Arabia; aaldalbahi@kfu.edu.sa; Tel.: +966-800-303-0308

Received: 6 September 2019; Accepted: 6 October 2019; Published: 10 October 2019

Abstract: In this paper, a novel link recover scheme is proposed for standalone (SA) millimeter wave communications. Once the main beam between the base station (BS) and the mobile station (MS) is blocked, then a bundle-beam is radiated that covers the spatial direction of the blocked beam. These beams are generated from an analog beamformer design that is composed of parallel adjacent antenna arrays to radiate multiple simultaneous beams, thus creating an analog beamformer of multiple beams. The proposed recovery scheme features instantaneous recovery times, without the need for beam scanning to search for alternative beam directions. Hence, the scheme features reduced recovery times and latencies, as opposed to existing methods.

Keywords: millimeter wave; analog beamforming; beam recovery; link blockage; multi-beam; recovery times

1. Introduction

Millimeter wave (mmWave) frequencies represents a major component of standalone (SA) 5G networks for high data rates support in enhanced mobile broadband (eMBB). One key advantage here is the contiguous available spectrum at these bands. However, the aggregated path losses impose the use of beamforming techniques to achieve higher link gains. The use of directional transmission and reception here (absence of omni-directional modes), requires the base station (BS) and mobile station (MS) to scan over all spatial directions to determine the best beamforming and combining vectors that yield the highest received signal level [1]. Consequently, this creates high computational complexity and prolonged access times, i.e., long control-plane latencies. This, in turn, contradicts with the International Mobile Telecommunications (IMT) framework requirements that define 10 milliseconds (ms) latency levels for eMBB in 5G systems [2]. Therefore, initial beam access schemes need to attain reduced times to achieve short control plane latencies. Consequently, beam access and adaptation arise as challenging problems in mmWave systems. Once initial access procedures are performed, the BS and MS need to maintain robust link adaptation when signal levels drop due to mobility and blockage effects. In particular, mmWave links are highly vulnerable to obstacles in the propagation paths between the MS and BS, which degrades signal levels and triggers link blockage [3]. This deficiency is more likely to occur when transmitting at narrow beams, i.e., short coherence times and low-channel ranks. Therefore, efficient beam recovery schemes are required to overcome link blockages, maintain communication sessions without drops, and reduce requirements for repeated beam access procedures.

In light of the above, this paper presents a novel beam recovery scheme to overcome link blockage effects and provides instantaneous beam recovery times, without the requirements for beam scanning. The scheme develops a bundle of simultaneously radiated beams that compensate for the blocked beam using an analog beamforming architecture. Namely, when blockage effects are introduced, the link is transiting from a line-of-sight (LoS) to a non-line-of-sight (NLoS) operation, which decays the signal quality. Conventional schemes require the beam to find alternative spatial directions by performing beam scanning. Meanwhile, the proposed scheme radiates the beam-bundle after blockage occurs, i.e., acting as backup beams. This, in turn, eliminates the need for beam scanning since the bundle radiates in different directions, hence resulting in signal aggregation at the receiver.

This paper is organized as follows: Firstly, Section 2 presents a survey on recent studies on beam recovery methods. Then, a proposed scheme is presented by first proposing the novel beamforming model in Section 3, along with the channel and signal models. This is followed by the bundle-beam recovery scheme in Section 4. Then, the performance evaluation is presented in Section 5, and finally, conclusions and future directions are discussed in Section 6.

2. Related Work

Multiple studies have investigated the beam recovery problem in mmWave communications. The work in [4,5] used hierarchical codebook-based procedures that restart the beam access search if a blockage effect is triggered. The authors in [6] presented a new beam aggregation method for fast beam recovery. The method utilized two beams to collectively add signal powers from the same direction. Moreover, an equal gain combining (EGC) scheme was presented in [7] that also combines multiple signals, which are received from the secondary and tertiary best directions, in order to overcome the signal losses caused by blockage at the main beam. Note that the aforementioned schemes are limited to low blockage parameters (obstacles of low density). Therefore, these schemes can yield unreliable links in dense scenarios.

Additionally, authors in [8] computed the signal level at the neighboring beam to the blocked direction affiliated with the main beam. Here the BS and MS are compelled to perform beam scanning, which features an increased number of measurements at the neighboring directions. In turn, this results in prolonged recovery times and vulnerability to communication sessions-drop. Moreover, a relay node method was presented in [9] that performs a handover decision when the direct link at the main beam is not recovered within a threshold time period. However, this scheme works only if at least triangulation geometry is available in the MS proximity. The work in [10] proposed a reactive beam recovery method, in which the MS exploits the microwave band to identify a back-up direction, in order to recover links from blockage without requirements for handover procedures. Note that the latter scheme is dependent on sub-6GHz microwave frequencies, which impedes the realization of SA mmWave networks.

3. System Model

3.1. Beamforming Design at the MS

Consider a MS equipped with an analog beamformer that is composed of parallel uniform linear arrays (ULA), where each ULA radiates a single beam, i.e., forming simultaneous multiple beams radiation in different directions. Each antenna is connected to a single analog phase shifter to provide continuous scanning capabilities (as opposed to step scanning in digital phase shifters). The ULAs are then connected to a single RF chain, as shown in Figure 1. Consider the design details.

In this paper, a novel multi-beam parallel array model is proposed at the MS. Hence, consider a MS equipped with a group of $r = 1, 2, \ldots, R$ parallel arrays, each composed of $n = 1, 2, \ldots, N_r$ co-polarized antenna elements arranged in a linear geometric setting, i.e., forming one-dimensional radiation (1D). The elements are uniformly oriented with d_r equi-spacing, i.e., $d = \lambda/2$, where λ represents the mmWave wavelength, $\lambda = c/f_c$, c is the speed of light, and f_c is the carrier frequency. This spacing value is chosen so that grating lobes and pattern blindness are avoided, as well as to ensure there are minimal mutual coupling effects. Thus, it satisfies the formula $d < \left(1 + |\cos\theta_0^r|\right)$, where the variable θ_0^r is the observation angle from array r at the MS in azimuth direction.

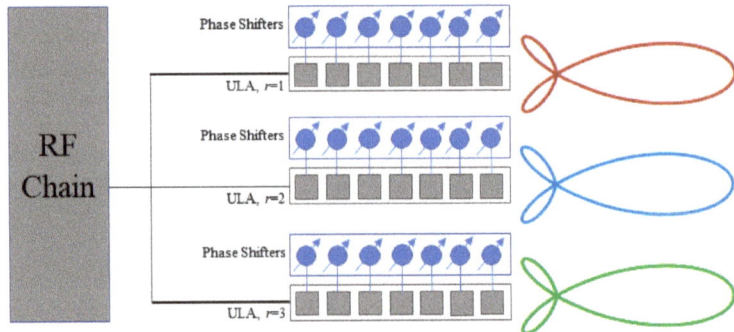

Figure 1. Proposed parallel analog beamformer.

The maximum radiation pattern for the array points along θ_0^r scanning directions, for an even number of elements at any spatial direction is expressed by the closed-form normalized array response vector at each array r at the MS; which, in turn, is represented by the periodic array factor (AF), A_{MS}^r. This is expressed as

$$A_{MS}^r = \frac{1}{N_{MS}^r} \sum_{n=1}^{N_{MS}^r} a_n exp(j(N_{MS}^r - 1)(k_v d_r \cos \theta_0^r + \beta_{MS}^r)) = A_{MS}^r = \frac{1}{N_{MS}^r} \sum_{n=1}^{N_{MS}^r} a_n exp(j(N_{MS}^r - 1)(\varphi_{MS}^r)) \quad (1)$$

where the variable a_n denotes the amplitude excitation for the n-th antenna element, $k_v = 2\pi/\lambda$ represents the wave number, and β_{MS}^r symbolizes the relative progressive phase shift between the interconnected antenna elements at array r at the MS. Note that $\varphi_{MS} = k_v d \cos \theta_0^{MS} + \beta_{MS}$ is a compact form that represents the array phase function at the MS with a visible region that varies between $-kd \leq \varphi_{MS} \leq kd$. Moreover, the half-power beamwidth (HPBW) at the broadside and scanning directions, i.e., $\forall \theta_0^{MS}(\theta_0^{MS}: 0 < \theta_0^{MS} \leq \pi)$, is expressed as [11]

$$\phi_{brd}^{MS} = \cos^{-1}\left(\frac{\lambda}{2\pi d} k_w d \cos \theta_0^{MS} \pm \frac{2.782}{N_{MS}^y}\right), 0 < \theta_0^{MS} \leq \pi, \quad (2)$$

whereas the HPBW at the endfire direction is computed as [12],

$$\phi_{endfire}^{MS} = 2\cos^{-1}(1 - 1.391\lambda/N_{MS}^y d), \; for \; \theta_0^{MS} = 0, \pi. \quad (3)$$

One important remark here, is that the spatial footprint of the array increases proportionally to a broadening factor of b, $b = 1/\cos \theta_0^{MS}$ for directions scanned off the broadside. Moreover, the array gain is gauged by $G_{A_{MS}} = A_{MS} G_a$, where G_a is the gain for a single antenna element. For example, microstrip rectangular patch antennas are widely chosen for mmWave transceivers, and they provide a gain range of 5–7 dBi [13,14].

Each antenna array is fed in parallel by an array of P phase shifters, in particular, quadrature varactor-loaded transmission-line phase shifters. Note that the total number of phase shifters is equal to the number of antennas. Varactor phase shifters are chosen here due to their high shifting times (in μs), low power requirement, reduced loss rates, and capability to continuously adjust and control the [0, 2π] spatial plane using a single control voltage unit. The phase shifters are then connected to a single RF chain. Overall, this structure formulates an analog beamfomer composed of multiple radiated beams that carry the same modulated data. The benefit of using an analog beamformer here is to reduce the power consumption levels associated with the RF chains, as in the case of digital and hybrid architectures. Since a single RF chain is used at the MS, the orthogonal beam coding technique is adopted here. Namely, the weights of the antenna elements are modified by a unique set of codes to produce unique beams of distinguished signals. The orthogonal codes here create distinguishable spatial signatures for each beam-bundle, and thereby can identify the exact direction of the highest received signal in the bundle-beam from that particular direction.

Hence, this work exploits orthogonal Hamming codewords, c_m, i.e., $c_m[e, d_H]$, where e is the codeword length, and d_H is the Hamming distance between successive codewords. Additionally, each codeword is scaled by the control signals, z, and features $[b_1, b_2, \ldots, b_F]$ codebits, where b_F is the total number of codebits. Consider the following codewords developed for a single bundle-beam, represented as:

$$\begin{array}{ll} c_1 = [-1\ -1\ -1\ -1\ -1\ -1\ -1\ -1] & c_2 = [1\ -1\ -1\ -1\ 1\ 1\ 1\ -1] \\ c_3 = [-1\ 1\ -1\ -1\ 1\ 1\ -1\ 1\ 1\ 1] & c_4 = [-1\ -1\ 1\ -1\ 1\ 1\ 1\ -1\ 1] \end{array} \quad (4)$$

These codewords feature zero cross-correlation, which yields in orthogonal beams in the bundle, hence receiving distinguishable signals in the directions of the beams [15,16]. Here, each codebit in the Hamming codeword is applied to the weight of a single antenna, where the number of codebits b_F is equal to the number of antennas N_r in the parallel array r. The codebit is either "1" or "−1". If it is "1", then the weight of the antenna remains the same, i.e., same amplitude and phase. Meanwhile, if the codeword is "−1", then the conjugate is applied to the weight, i.e., keeping the same amplitude and rotating the phase by π.

These codes are reciprocal at the MS and BS. Therefore, when a signal is received at the BS, it is basically receiving one codeword. Namely, it multiplies the received codeword (appearing in the weights of the antennas) by all the four codewords, in order to retrieve the unique codeword and its affiliated beam in the bundle. As a result, the BS now identifies the directions with the highest signal level.

3.2. Digital Beamformer at the BS

Digital beamforming is adopted at the BS due to the abundant power input levels, and the necessity to provide multi-user connectivity. The beamformer architecture shown in Figure 2 is based upon a uniform circular array (UCA) with an identical radiation pattern of symmetric beamwidth in all spatial directions (no beam broadening at endfire direction), i.e., providing similar signal levels to MSs at different locations, with high directivities. The UCA here also features reduced sidelobe levels (SLL), and it eliminates the need for back-to-back arrays, as the case in the 1D ULA. Hence, consider a BS equipped with a UCA composed of N_{BS} total number of antenna elements, which are uniformly spaced on the x–y plane along radius, a, in a circular geometric setting. Now each n_{BS} antenna, $n_{BS} \in N_{BS}$, is also connected to an analog phase shifter to provide continuous beam scanning. This structure is connected to a group of RF chains R_{BS}, where the total number of RF chains is equal to the number of antennas. Overall, this setting results in a single beam radiated from each antenna, i.e., represented by the beamforming vector, v_{BS}, $v_{BS} \in V_{BS}$; where V_{BS} is the beamforming matrix that represents the beam-bundle at the BS, B_{BS}, such that $V_{BS} = V_{bb} V_{an}$, where V_{bb} and V_{an} denote, in order, the beamforming matrices at the baseband and analog stages, i.e., $V_{bb} = n_{BS} \times r_{BS}$ and $V_{an} = r_{BS} \times n_{BS}$. Here, each vector v_{BS} carries unique modulated data that can be utilized for multi-users, or it supports a single datum to support MS with a beam-bundle when blockage occurs. This vector is gauged by the AF for the UCA, A_{BS} i.e.,

$$A_{BS} = \sum_{n=1}^{N_{BS}} I_n \exp(jva \sin\theta \cos(\phi - \phi_n^{BS}) + \varphi_s^{BS}, \quad (5)$$

where the angles θ and ϕ represent the directions along the y- and x- axes, φ_n^{BS} is the angular position of the n-th antenna, where $\varphi_n^{BS} = 2\pi n_{BS}/N_{BS}$. Moreover, the variables I_n, v and φ_s^{MS} in Equation (5) symbolizes the amplitude of the n-th antenna, the wave number, and the maximum radiation principal at the BS, evaluated as

$$\varphi_0^{BS} = -va \sin \theta_0^{BS} \cos(\phi_0^{BS} - \phi_n^{BS}) \qquad (6)$$

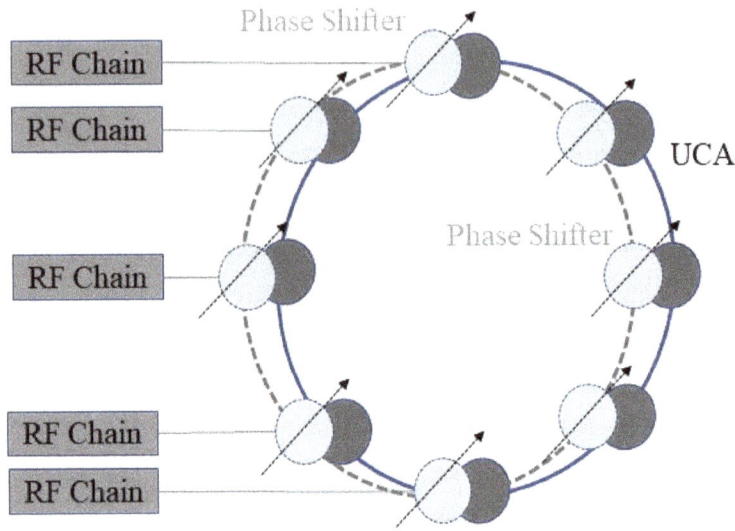

Figure 2. Digital beamformer at the BS.

3.3. Signal Model

Consider that MS and BS entities operate in LoS settings in urban outdoor environments composed of various objects in the proximity of the MS. In addition, assume a full-duplex division duplexing (FDD) channel of reciprocal channel state information (CSI) at both entities. Then, the downlink (DL) received signal profile at the MS, y_{an}, is expressed as

$$Y_{bb} = P_{tr} U_{MS}^H V_{BS} \mathbf{H} z + U_{MS}^H w, \qquad (7)$$

where P_{tr}, $\mathbf{H}$, z, and w denote in order the transmitted signal power, the power complex channel, the reference control signal, and the additive white Gaussian noise (AWGN), i.e., $w \sim N(0, \sigma_w^2)$, where σ_w^2 is the noise variance.

3.4. Channel Model

The geometric channel model, $\mathbf{H}$, is adopted here due to the scattering nature of mmWave propagation. This is highly attributed to the large obstacle dimensions, as compared to the propagating wavelength at these bands. Consequently, this yields in reduced scattering profile, and hence results in poor scattering signal profile of a low number of rays, i.e., Poisson distribution. In turn, this results in high dependence on the geometry of the objects in the propagation link. This model is expressed as [17]

$$\mathbf{H} = \sqrt{\frac{N_{BS} N_{MS}}{\Gamma_{bl}}} \sum_{k=1}^{K} \sum_{l=1}^{L} h_l V_{BS} U_{MS}^H, \qquad (8)$$

where the variables Γ_{bl}, and h_l represent the blockage path loss model, and the gain of the l-th path. The signal profile here, is composed of L total number of paths that are observed in K total number of clusters, i.e., $L \in K$. These paths follow Rician-distribution that accounts for the LoS-to-NLoS plink transition caused by blockage effects. Namely, the path gain is modeled as $h_l \sim R(0, \zeta)$, where ζ is the power ratio between the dominant and other paths. Moreover, the beamforming and combining matrices, V_{BS} and U_{MS} (which also represent the response vectors), are evaluated using their far-field array factors (AF), as presented in the beamforming models.

3.5. Blockage Model

As mentioned earlier, the blockage path loss model, Γ_{bl}, accounts for LoS-to-NLoS link transition, when obstacles of different densities are present in the direct propagation link affiliated with the main beam. This model is formulated as [18]

$$\Gamma_{bl} = \mathbb{I}[\mathbb{P}(d)]\Gamma_{LoS}(d) + \mathbb{I}[1 - \mathbb{P}(d)]\Gamma_{NLoS}(d), \tag{9}$$

where $\mathbb{I}$ is an indicator function that specifies the link-blockage state, i.e., $\mathbb{I}(x) = 1$ iff $x = 1$, and it is set as $\mathbb{I}(x) = 0$ otherwise. Moreover, the variables $\Gamma_{LoS}(d)$ and $\Gamma_{NLoS}(d)$ represent the path loss for the LoS and NLoS settings, respectively, expressed as [19],

$$\Gamma_{LoS}(d) = 10\log_{10}(d_{ref}) + 10\,\delta_{LoS}\log_{10}(d),\ for\ LoS, \tag{10}$$

$$\Gamma_{NLoS}(d) = 10\log_{10}(d_{ref}) + 10\,\delta_{NLoS}\log_{10}(d),\ for\ NLoS, \tag{11}$$

where the variable d represents the distance between the BS and MS, d_{ref} is close-in reference distance, and δ_{LoS} and δ_{NLoS} are the path loss exponents (PLE) for the LoS and NLoS settings, respectively. Moreover, the notations $\mathbb{P}(d)$ and $(1-\mathbb{P}(d))$ denote the LoS and NLoS probabilities at the distance d. Here the probability is $\mathbb{P}(d) = exp(-\rho d)$, where ρ is the blockage parameter that accounts for obstacles of different dimensions and densities. Note that the higher the blockage parameter, the more blockage effects are caused to the link.

4. Recovery Procedure

Consider that MS and BS entities operate in a SA mmWave network in LoS settings of Rician scattering. During the initial access stage, an iterative random search is conducted over all spatial directions at the MS and B. This process returns the best beamforming and combining vectors (best pointing directions) that yield the highest received signal level, modeled as

$$(u_{MS}, v_{BS})_{bst} = max(\mathcal{Y}_{u,v}). \tag{12}$$

where these best vectors $(u_{MS}, v_{BS})_{bst}$ present the maximum principal directions of the primary beams at the MS and BS, which are selected for the data-plane transmission. Now, once the session starts, the spectral efficiency can take various levels based on the link quality. First, when the link is in LoS, it features high link quality without obstacles (blockage parameter is zero), $\mathbb{I}(x) = 0$, as well as high instantaneous spectral efficiency, δ_{inst}.

When the obstacles in the propagating path become present in the direct link associated with the main beam, it starts to exhibit instantaneous low spectral efficiency, and then blockage mode is in effect. Here the indicator function is set as $\mathbb{I}(x) = 1$, to indicate the LoS-to-NLoS transition. The blockage threshold is set based on the spectral efficient level, as

$$\delta_{inst} < min\{\log_2(1 + 10^{0.1(SNR-\Omega)}), \delta_{max}\}, \tag{13}$$

where SNR stands for the signal-to-noise ratio, the variable Ω denotes the loss factor (measured in dB), and δ_{max} represents the maximum spectral efficiency [20]. Note that the SNR is expressed as,

$$SNR = \frac{P_{tr}G_{MS}G_{BS}|h_l|^2}{\Psi T_0 \bar{\sigma}}, \tag{14}$$

where Ψ, denotes the Boltzmann constant, T_0 is the operating temperature, $\bar{o}$ is the channel bandwidth, and G_{MS} and G_{BS} are the array gains at the MS and BS respectively. $G_{MS} = g_n U_{MS}$, and $G_{BS} = g_n V_{BS}$, where g_n is the gain for a single antenna element.

In light of the above, when the main beam is blocked, the MS and BS initiate the beam-bundle as the backup beams to compensate for the signal losses associated with the main beam, see Figure 3. Therefore, session drops are avoided. The MS here performs maximal ratio combining (MRC) to amplify high beam signals and attenuate weak beam signals, i.e.,

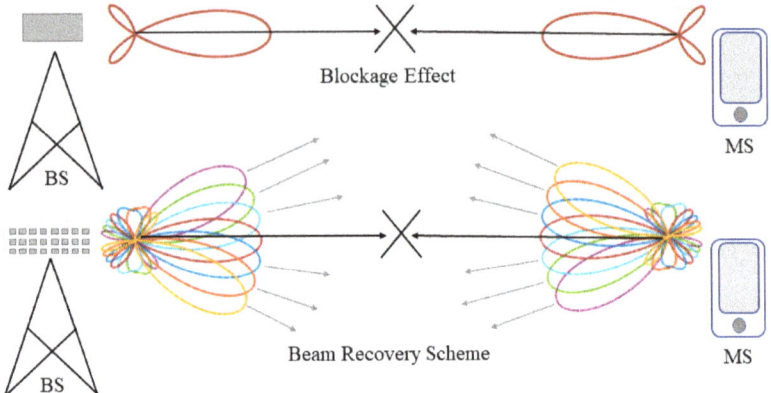

Figure 3. Proposed multi-backup beam recovery scheme.

5. Performance Evaluation

The proposed recovery scheme is now evaluated using key metrics, in particular, the spectral efficiency, received signal profile, and recovery times. Consider Table 1 for the overall system settings.

Table 1. System parameters.

Category	Parameter	Value
System	f_α (GHz), b (MHz), P_{tr}(dBm)	28, 700, 30
Arrays	a, g_n, N_{MS}, N_{BS},	1, 5, 256, 256
Channel	σ_w^2 (dB) ζ, d (dB), ρ	1, 3, 5, 250, 0–1
Path loss	d_{ref} (m), δ_{LoS}, δ_{NLoS}	5, 2.6, 4
Spectral efficiency	Ω, δ_{max}	2, 10

5.1. Spectral Efficiency

Figure 4 shows the spectral efficiency for the proposed scheme at various blockage densities. The proposed scheme aims to enhance the spectral efficiency once beam blockage is in effect. Figure 4 shows the spectral efficiency for the proposed scheme versus conventional recovery methods that test neighboring beam directions or reset beam scanning procedures. When the direct propagation link between the MS and BS is free of obstacles (LoS operation), i.e., $\mathbb{I}(x) = 0$, high spectral efficiency is observed here. As a result, conventional methods and the proposed bundle-beam scheme yield high spectral efficiency. However, when obstacles start to appear in the propagation path of the primary beam, then the received signal level degrades, affecting capacity levels, and thereby reducing the spectral efficiency, as observed for the conventional schemes. For example, the work in [8] testing neighboring beams, requires beam scanning. Moreover, the beam aggregation method enhances the spectral efficiency if the density of the obstacles is low. However, this method results in a small number of scattering paths, attributed to the limited spatial coverage provided by the recovery narrow beams, which has a HPBW that is smaller than the dimensions of the obstacles. Meanwhile, the proposed scheme achieves high spectral efficiency if blockage parameters are dense. This is attributed to the wide spatial space covered by the back-up beam-bundle, which also yields in a high scattering profile that is leveraged using MRC.

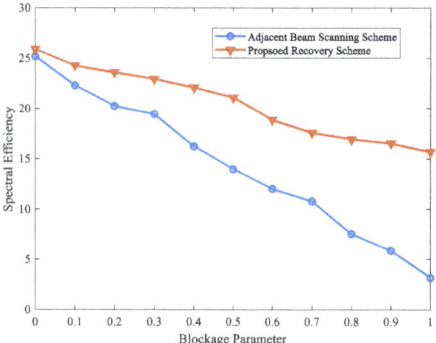

Figure 4. Spectral efficiency for various blockage parameters.

5.2. Received Power Delay Profile

Figure 5 shows the received power profile in time domain for the proposed scheme as opposed to conventional schemes (i.e., testing adjacent beam directions) shown in Figure 6. The proposed recovery scheme yields a rich scattering profile due to the wide spatial coverage achieved by the instantaneous back-up beams, with HPBW that exceeds the obstacle's dimensions, as well as enriching the reflections in the Rician path gains in the channel settings. For example, the proposed scheme exhibits 4–5 rays in 2–3 clusters when blockage is triggered, as opposed to 2–3 rays in 1–2 clusters for the neighboring beams testing and conventional codebook schemes. Furthermore, the recorded clusters here are received with power levels of −60 dBm, which relaxes the receiver sensitivity requirements. This is compared to −120 dBm and −140 dBm signal levels for the other schemes, which results in significant challenges in acquiring the signal, liming coverage ranges, as well as providing poor low channel capacity and impeding high channelization services for SA mmWave networks.

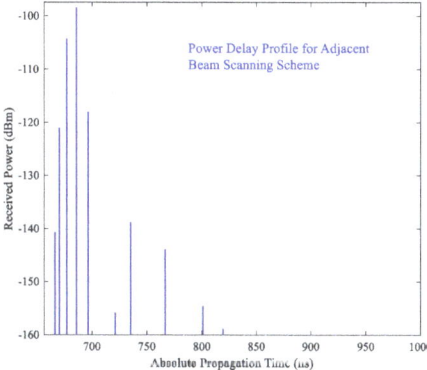

Figure 5. Power delay profile for conventional schemes (adjacent beam and aggregation).

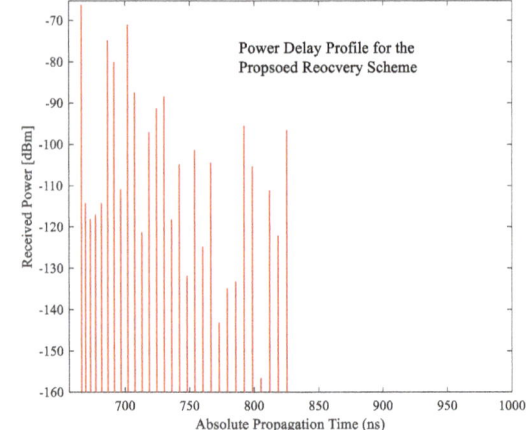

Figure 6. Power delay profile for proposed multi-backup scheme.

5.3. Beam Recovery Times

One major performance metric for beam recovery schemes is the beam recovery time. It is defined as the overall time period that is required to determine an alternative link direction once the direct link of the main beam is blocked. Namely, it is the period required to determine the new best beamforming and combining vectors, and their principal directions at the MS and BS. Figure 7 shows the recovery times at the MS T_{rec}^{MS} (likewise for the BS, T_{rec}^{BS}). The proposed scheme achieves instantaneous recovery times without the requirement for beam scanning or resetting the access schemes when a link is blocked. The only time required here is the PSS transmission duration of the beam vectors (i.e., 200 μs). Overall, the scheme here promotes the feasibility of SA mmWaves with ultra-low recovery times, thereby realizing low latency requirements.

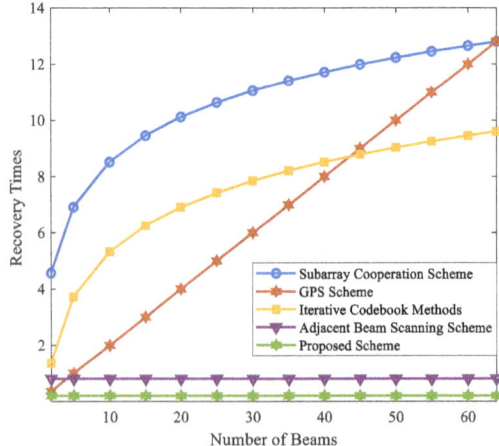

Figure 7. Recovery time for the proposed multi-backup scheme at various beam numbers.

6. Conclusions

This paper presents a novel beam recovery scheme for standalone millimeter wave communications, without the reliance on sub 6 GHz microwave bands assistance. The scheme is based on novel analog beamforming architecture that radiates multiple instantaneous beam directions from a single RF chain.

Here, once the main beam is blocked, then a group of adjacent simultaneous beams are radiated that cover the blocked beam and promotes maximum ratio combining at the MS for enhanced signal profile and spectral efficiency. The proposed scheme features instantaneous recovery times, which can realize ultra-low latency levels. Future efforts will investigate the effects of user mobility, coherence times, and terrain in blockage modeling, and evaluating the proposed scheme under these effects.

Funding: This work is funded by the Deanship of Scientific Research at King Faisal University, under Nasher Track (grant number 186239).

Acknowledgments: The author acknowledges the Deanship of Scientific Research at King Faisal University for their continuous support during this project.

Conflicts of Interest: The author declares no conflict of interest.

References

1. Liu, J. Initial Access, Mobility, and User-Centric Multi-Beam Operation in 5G New Radio. *IEEE Commun. Mag.* **2018**, *56*, 35–41. [CrossRef]
2. International Mobile Telecommunications. *Minimum Requirements Related to Technical Performance for IMT-2020 Radio Interface*; ITU-R Study Group 5: Geneva, Switzerland, 2017; pp. 6–7. Available online: https://www.itu.int/pub/R-REP-M.2410-2017 (accessed on 4 June 2019).
3. Huo, Y.; Dong, X.; Xu, W. 5G Cellular User Equipment: From Theory to Practical Hardware Design. *IEEE Access* **2017**, *5*, 13992–14010. [CrossRef]
4. Jasim, M.; Aldalbahi, A.; Khreishah, A.; Ghani, N. Hooke Jeeves Search Method for Initial Beam Access in 5G mmWave Cellular Networks. In Proceedings of the 2017 IEEE 28th Annual International Symposium on Personal, Indoor, and Mobile Radio Communications (PIMRC), Montreal, QC, Canada, 8–13 October 2017.
5. Jasim, M.; Ghani, N. Generalized Pattern Search for Beam Discovery in Millimeter Wave Systems. In Proceedings of the 2017 IEEE 86th Vehicular Technology Conference (VTC-Fall), Toronto, ON, Canada, 24–27 September 2017.
6. Jasim, M.; Aldalbahi, A.; Shakhatreh, H. Beam Aggregation for Instantaneous Link Recovery in Millimeter Wave Communications. In Proceedings of the IEEE International Conference on Wireless and Mobile Computing, Networking and Communications (WIMOB), Limassol, Cyprus, 15–17 October 2018.
7. Jasim, M.; Ababneh, M.; Siasi, N.; Ghani, N. Hybrid Beamforming for Link Recovery in Millimeter Wave Communications. In Proceedings of the 2018 IEEE Wireless and Microwave Technology Conference (WAMICON), Clearwater, FL, USA, 15–17 April 2018.
8. Gao, B.; Xiao, Z.; Zhang, C.; Su, L.; Jin, D.; Zeng, L. Double-Link Beam Tracking Against Human Blockage and Device Mobility for 60-GHz WLAN. In Proceedings of the 2014 IEEE Wireless Communications and Networking Conference (WCNC), Istanbul, Turkey, 6–9 April 2014.
9. Kim, W.; Song, J.; Baek, S. Relay-Assisted Handover to Overcome Blockage in Millimeter-Wave Networks. In Proceedings of the 2017 IEEE 28th Annual International Symposium on Personal, Indoor, and Mobile Radio Communications (PIMRC), Montreal, QC, Canada, 8–13 October 2017.
10. Giordani, M.; Mezzavilla, M.; Rangan, S.; Zorzi, M. An Efficient Uplink Multi-Connectivity Scheme for 5G Millimeter-Wave Control Plane Applications. *IEEE TWC* **2018**, *17*, 6806–6821. [CrossRef]
11. Balanis, C. Arrays: Linear, Planar and Circular. In *Antenna Theory: Analysis and Design*, 3rd ed.; John Wiley & Sons: Hoboken, NJ, USA, 2005; pp. 290–304.
12. Stutzman, W.; Thiele, G. Array Antennas. In *Antenna Theory and Design*, 3rd ed.; John Wiley & Sons: Hoboken, NJ, USA, 2012; pp. 626–629.
13. Firdausi, A.; Alaydrus, M. Designing multiband multilayered microstrip antenna for mmWave applications. In Proceedings of the 2016 International Conference on Radar, Antenna, Microwave, Electronics, and Telecommunications (ICRAMET), Jakarta, Indonesia, 3–5 October 2016.
14. Altaf, A.; Alsunaidi, M.A.; Arvas, E. A novel EBG structure to improve isolation in MIMO antenna. In Proceedings of the 2017 USNC-URSI Radio Science Meeting (Joint with AP-S Symposium), San Diego, CA, USA, 9–14 July 2017.

15. Tsang, Y.; Poon, A.; Addepalli, S. Coding the Beams: Improving Beamforming Training in mmWave Communication System. In Proceedings of the 2011 IEEE Global Telecommunications Conference—GLOBECOM 2011, Houston, TX, USA, 5–9 December 2011.
16. Sur, S.; Zhang, X.; Ramanathan, P.; Chandram, R. BeamSpy: Enabling Robust 60 GHz Links Under Blockage. In Proceedings of the 2016 USENIX Workshop on Cool Topics in Sustainable Data Centers, Santa Clara, CA, USA, 16 March 2016.
17. Alkhateeb, A.; El Ayach, O.; Leus, G.; Heath, R. Channel Estimation and Hybrid Precoding for Millimeter Wave Cellular Systems. *IEEE J. Sel. Top. Signal Process.* **2014**, *8*, 831–846. [CrossRef]
18. Bai, T.; Desai, V.; Heath, R. Millimeter Wave Cellular Channel Models for System Evaluation. In Proceedings of the 2014 International Conference on Computing, Networking and Communications (ICNC), Honolulu, HI, USA, 3–6 February 2014.
19. Sun, S.; Rappaport, T.S.; Rangan, S.; Thomas, T.A.; Ghosh, A.; Kovacs, I.Z.; Rodriguez, I.; Koymen, O.; Partyka, A.; Jarvelainen, J. Propagation Path Loss Models for 5G Urban Micro- and Macro-Cellular Scenarios. In Proceedings of the 2016 IEEE 83rd Vehicular Technology Conference (VTC2016-Spring), Nanjing, China, 15–18 May 2016.
20. Akdeniz, M.R.; Liu, Y.; Samimi, M.K.; Sun, S.; Rangan, S.; Rappaport, T.S.; Erkip, E. Millimeter Wave Channel Modeling and Cellular Capacity Evaluation. *IEEE J. Sel. Areas Commun.* **2014**, *32*, 1164–1179. [CrossRef]

© 2019 by the author. Licensee MDPI, Basel, Switzerland. This article is an open access article distributed under the terms and conditions of the Creative Commons Attribution (CC BY) license (http://creativecommons.org/licenses/by/4.0/).

Article

Correlated Blocking in mmWave Cellular Networks: Macrodiversity, Outage, and Interference

Enass Hriba and Matthew C. Valenti *

Lane Department of Computer Science and Electrical Engineering, West Virginia University, Morgantown, WV 26506, USA; efhriba@mix.wvu.edu
* Correspondence: valenti@ieee.org

Received: 30 September 2019; Accepted: 16 October 2019; Published: 18 October 2019

Abstract: In this paper, we provide a comprehensive analysis of macrodiversity for millimeter wave (mmWave) cellular networks. The key issue with mmWave networks is that signals are prone to blocking by objects in the environment, which causes paths to go from line-of-sight (LOS) to non-LOS (NLOS). We identify macrodiversity as an important strategy for mitigating blocking, as with macrodiversity the user will attempt to connect with two or more base stations. Diversity is achieved because if the closest base station is blocked, then the next base station might still be unblocked. However, since it is possible for a single blockage to simultaneously block the paths to two base stations, the issue of correlated blocking must be taken into account by the analysis. Our analysis characterizes the macrodiversity gain in the presence of correlated random blocking and interference. To do so, we develop a framework to determine distributions for the LOS probability, Signal to Noise Ratio (SNR), and Signal to Interference and Noise Ratio (SINR) by taking into account correlated blocking. We validate our framework by comparing our analysis, which models blockages using a random point process, with an analysis that uses real-world data to account for blockage. We consider a cellular uplink with both diversity combining and selection combining schemes. We also study the impact of blockage size and blockage density along with the effect of co-channel interference arising from other cells. We show that the assumption of independent blocking can lead to an incorrect evaluation of macrodiversity gain, as the correlation tends to decrease macrodiversity gain.

Keywords: correlated blocking; millimeter wave; line-of-sight; macrodiversity

1. Introduction

Millimeter-wave (mmWave) has emerged in recent years as a viable candidate for infrastructure-based (i.e., cellular) systems [1–5]. Communicating at mmWave frequencies is attractive due to the potential to support high data rates at low latency [1,2,6]. At mmWave frequencies, signals are prone to blocking by objects intersecting the paths and severely reducing the signal strength, and thus the Signal to Noise Ratio (SNR) [7–10]. For instance, blocking by walls provides isolation between indoor and outdoor environments, making it difficult for an outdoor base station to provide coverage indoors [11]. To mitigate the issue of blocking in mmWave cellular networks, macrodiversity has emerged as a promising solution, where the user attempts to connect to multiple base stations [12]. With macrodiversity, the probability of having at least one line-of-sight (LOS) path to a base station increases, which can improve the system performance [13–15].

An effective methodology to study wireless systems in general, and mmWave systems in particular, is to embrace the tools of stochastic geometry to analyze the SNR and interference in the network [3,15–20]. With stochastic geometry, the locations of base stations and blockages are assumed to be drawn from an appropriate point process, such as a Poisson point process (PPP). When blocking is modeled as a random process, the probability that a link is LOS is an exponentially decaying function of link distance. While many papers assume that blocking is independent [11,17], in reality the blocking of multiple paths may be correlated [18]. The correlation effects are especially important for macrodiverity networks when base stations are close to each other, or more generally when base stations have a similar angle to the transmitter. In this case, when one base station is blocked, there is a significant probability that another base station is also blocked [13–15].

Prior work has considered the SINR distribution of mmWave personal networks [16,17,21]. Such work assumes that the blockages are drawn from a point process (or, more specifically, that the centers of the blockages are drawn from a point process and each blockage is characterized by either a constant or random width). Meanwhile, the transmitters are either in fixed locations or their locations are also drawn from a point process. A universal assumption in this prior art is that the blocking is independent; i.e., each transmitter is blocked independently from the other transmitters. As blocking has a major influence on the distribution of signals, it must be carefully taken into account. Independent blocking is a crude approximation that fails to accurately capture the true environment, especially when the base stations, or, alternatively, the user equipments (UEs), are closely spaced in the angular domain or when there are few sources of blocking. We note that blocking can be correlated even when the sources of blockage are placed independently according to a point process.

The issue of blockage correlation was considered in [22–25], but it was in the context of a localization application where the goal was to ensure that a minimum number of positioning transmitters were visible by the receiver. As such, this prior work was only concerned with the *number* of unblocked transmissions rather than the distribution of the received aggregate signal (i.e., source or interference power). In [18], correlated blocking between interferers was considered for wireless personal area network. Recently, correlation between base stations was considered in [13,14] for infrastructure-based networks with macrodiversity, but in these references the only performance metric considered is the nth order LOS probability; i.e., the probability that at least one of the n closest base stations is LOS. However, a full characterization of performance requires other important performance metrics, including the distributions of the SNR and, when there is interference, the Signal to Interference and Noise Ratio (SINR). Alternatively, the performance can be characterized by the coverage probability, which is the complimentary cumulative distribution function of the SNR or SINR, or the rate distribution, which can be found by using information theory to link the SNR or SINR to the achievable rate.

In this paper, we propose a novel approach for fully characterizing the performance of macrodiversity in the presence of correlated blocking. While, like [13,14], we are able to characterize the spatially averaged LOS probability (i.e., the LOS probability averaged over many network realizations), our analysis shows the *distribution* of the LOS probability, which is the fraction of network realizations that can guarantee a threshold LOS probability rather than its mere spatial average. Moreover, we are able to similarly capture the distributions of the SNR and SINR. Furthermore we validate our framework by comparing the analysis to a real data building model.

We assume that the centers of the blockages are placed according to a PPP. We first analyze the distributions of LOS probability for first- and second-order macrodiversity. We then consider the distribution of SNR and SINR for the cellular uplink with both selection combining and diversity combining. The signal model is such that blocked signals are completely attenuated, while LOS, i.e., non-blocked, signals are subject to an exponential path loss and additive white Gaussian noise (AWGN). Though it complicates the exposition and notation, the methodology can be extended to more elaborate models, such as one wherein all signals are subject to fading and non-LOS (NLOS) signals are partially attenuated (see, e.g., [17]).

The remainder of the paper is organized as follows. We begin by providing the system model in Section 2, wherein there are base stations and blockages, each drawn from a PPP. In Section 3 we provide an analysis of the LOS probability under correlated blocking and derive the blockage correlation coefficient using arguments based on the geometry and the properties of the blockage point process; i.e., by using stochastic geometry. Section 4 provides a framework of the distribution of SNR, where the results depend on the blockage correlation coefficient. In Section 5, we validate our framework by comparing the analysis to a real data model. Then in Section 6, interference is considered and the SINR distribution is formalized. Finally, Section 7 concludes the paper, suggesting extensions and generalizations of the work.

2. System Model

2.1. Network Topology

Consider a mmWave cellular network consisting of base stations, blockages, and a source transmitter, which is a UE. The UE attempts to connect to the N closest base stations, and therefore operates in a Nth order macrodiversity mode. The locations of the base stations are modeled as an infinite homogeneous PPP with density λ_{bs}. We assume the centers of the blockages also form a homogeneous PPP with density λ_{bl}, independent from the base station process. Let Y_0 indicate the source transmitter and its location. Due to the stationarity of the PPPs, and without loss of generality, we can assume the coordinates are chosen such that the source transmitter is located at the origin; i.e., $Y_0 = 0$. In Section 6, we will consider additional transmitters located in neighboring cells, which act as interferers.

Let X_i for $i \in \mathbb{Z}^+$ denote the base stations and their locations. Let $R_i = |X_i|$ be the distance from Y_0 to X_i. Base stations are ordered according to their distances to Y_0 such that $R_1 \leq R_2 \leq \ldots$. The signal of the source transmitter is received at the closest N base stations, and hence, N is the number of X_i connected to Y_0. For a PPP, a derivation of the distribution of $R_1, \ldots, R_N$ is given in Appendix B, which implies a methodology for generating these distances within a simulation.

Figure 1 shows an example of second-order macrodiversity ($N = 2$) cellular network where the user attempts to connect to its closest two base stations. The solid line indicates the link from the user to the base station is LOS, while the dashed line indicates the link is NLOS. The figure shows examples of two different blockage scenarios. In Figure 1a the closest base station (X_1) is LOS while X_2 is NLOS to the user, in which case the blockage only blocks a single link. In Figure 1b a single blockage blocks both links to X_1 and X_2. The fact that sometimes a single blockage can block both links is an illustration of the effect of correlated blocking.

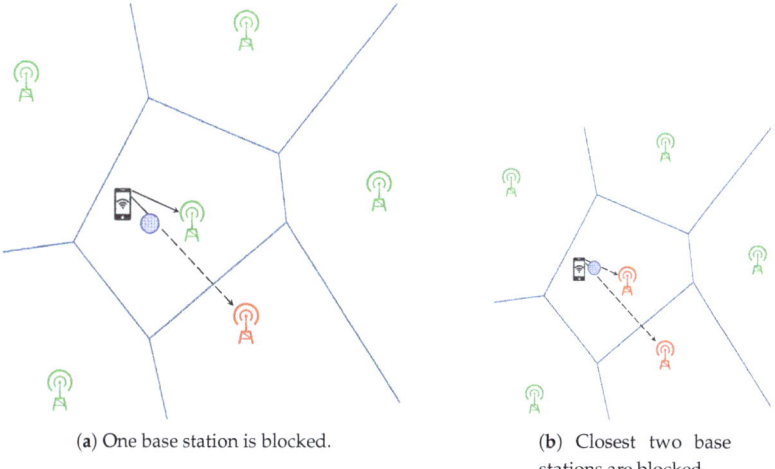

(a) One base station is blocked.

(b) Closest two base stations are blocked.

Figure 1. Example network topology with two different blockage scenarios. The source transmitter is the mobile device shown in the central cell. Its signal is transmitted to its closest two base stations. The solid line indicates the link is line-of-sight (LOS), while the dashed line indicates the link is non-LOS (NLOS).

2.2. Blockage Model

As in [18], each blockage is a segment of length W. To capture the worst-case scenario, as shown in Figure 2a, it is assumed that the line representing the blockage is perpendicular to the line that connects it to the transmitter. Although W can itself be random as in [13], we assume here that all blockages have the same value of W. In Figure 2a, the red stars indicate the blocked base stations, which are located in the blue shaded region. If a blockage cuts the path from Y_0 to X_i, then the signal from Y_0 is NLOS, while otherwise it is LOS. Here, we assume that NLOS signals are completely blocked while LOS signals experience exponential path-loss with a path-loss exponent α; i.e., the power received by X_i is proportional to $R_i^{-\alpha}$.

Each base station has a blockage region associated with it, illustrated by the blue shaded rectangles shown in Figure 2b. We use a_i to denote the blockage region associated with X_i and its area; i.e., a_i is both a region and an area. If the center of a blockage falls within a_i, then X_i will be blocked since at least some part of the blockage will intersect the path between X_i and Y_0. Because a_i is a rectangle of length R_i and width W, it is clear that $a_i = WR_i$. Unless X_1 and X_2 are exactly on opposite sides of the region, there will be an overlapping region v common to both a_1 and a_2. Because of the overlap, it is possible for a single blockage to simultaneously block both X_1 and X_2 if the blockage falls within region v, which corresponds to correlated blocking.

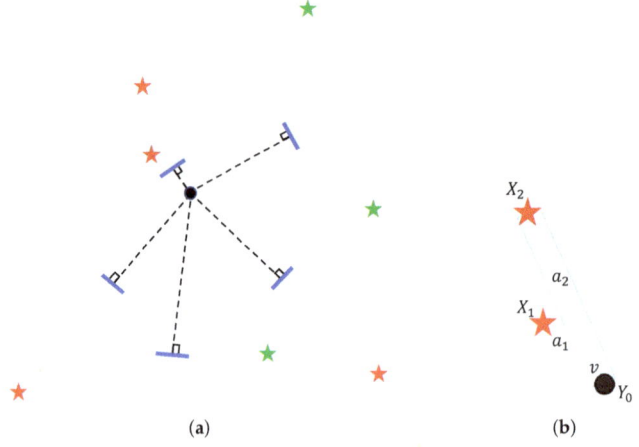

Figure 2. Illustration of the blockage model. (**a**) Network example consisting of base stations indicated by the stars and blockages indicated by blue lines surrounding the transmitter, which is indicated by the black circle. The blockages are modeled as a line of length W facing the transmitter; (**b**) Equivalent blockage regions. a_1 and a_2 are the blockage areas, and v is the overlapping area.

3. LOS Probability Analysis Under Correlated Blocking

In this section, we analyze the LOS probability, which is denoted p_{LOS}, and the impact of blockage correlation. Our focus is on second-order macrodiversity, where the signal of the source transmitter Y_0 is received at the two closest base stations X_1 and X_2. The LOS probability is the probability that at least one X_i is LOS to the transmitter. Because the base stations are randomly located, the value of p_{LOS} will vary from one network realization to the next, or equivalently by a change of coordinates, from one source transmitter location to the next. Hence, p_{LOS} is itself a random variable and must be described by a distribution. To determine p_{LOS} and its distribution, we first need to define the variable B_i which indicates that the path between Y_0 and X_i is blocked. Let $p_{B_1,B_2}(b_1,b_2)$ be the joint probability mass function (pmf) of $\{B_1, B_2\}$. Let p_i denote the probability that $B_i = 1$, which indicates the link from Y_0 to X_i is NLOS. Furthermore, let $q_i = 1 - p_i$, which is the probability that the link is

LOS, and ρ denote the correlation coefficient between B_1 and B_2. As shown in Appendix A, the joint pmf of $\{B_1, B_2\}$ as a function of ρ found to be

$$p_{B_1,B_2}(b_1, b_2) = \begin{cases} q_1 q_2 + \rho h & \text{for } b_1 = 0, b_2 = 0 \\ q_1 p_2 - \rho h & \text{for } b_1 = 0, b_2 = 1 \\ p_1 q_2 - \rho h & \text{for } b_1 = 1, b_2 = 0 \\ p_1 p_2 + \rho h & \text{for } b_1 = 1, b_2 = 1 \end{cases} \quad (1)$$

where $h = \sqrt{p_1 p_2 q_1 q_2}$.

For a two-dimensional homogeneous PPP with density λ, the number of points within an area a is Poisson with mean λa [26]. From the probability mass function of a Poisson variable, the probability of k points within the area is given by [27]

$$p_K(k) = \frac{(\lambda a)^k}{k!} e^{-\lambda a} \quad (2)$$

The event that the path to X_i is not blocked (LOS) by an object falling in area a_i can be obtained by the void probability of PPP, which is the probability that there are no blockages located in a_i, or equivalently, the probability that $k = 0$. Thus, q_i, which is equal to the void probability, is given by substituting $k = 0$ into (2) with $\lambda = \lambda_{bl}$ and $a = a_i$, which results in

$$q_i = \exp(-\lambda_{bl} a_i) \quad (3)$$

For first-order macrodiversity ($N = 1$), the LOS probability is given by q_1. Conversely, X_i will be NLOS when at least one blockage lands in a_i and this occurs with probability $p_i = 1 - q_i$ given by

$$p_i = 1 - \exp(-\lambda_{bl} a_i) \quad (4)$$

For second-order macrodiversity (N = 2), there will be a LOS signal as long as both paths are not blocked. This corresponds to the case that B_1 and B_2 are both not equal to unity. When blocking is not correlated, the corresponding LOS probability is $1 - p_1 p_2$. Correlated blocking may be taken into account by using (1) and noting that the LOS probability is the probability that B_1 and B_2 are not both equal to one, which is given by

$$p_{LOS} = 1 - p_{B_1,B_2}(1,1) = 1 - p_1 p_2 - \rho h \quad (5)$$

The blockage correlation coefficient ρ can be found from (1),

$$\rho = \frac{p_{B_1,B_2}(0,0) - q_1 q_2}{h} \quad (6)$$

where $p_{B_1,B_2}(0,0)$ is the probability that both X_1 and X_2 are LOS. Looking at Figure 2b, this can occur when there are no blockages inside a_1 and a_2. Taking into account the overlap v, this probability is the void probability for area $(a_1 + a_2 - v)$, which is given by

$$p_{B_1,B_2}(0,0) = e^{-\lambda_{bl}(a_1 + a_2 - v)} \quad (7)$$

Details on how to compute the overlapping area v are provided in [18]. Substituting (7) into (6) into (5) and using the definitions of p_i and q_i yields

$$p_{LOS} = e^{-\lambda_{bl} a_1} + e^{-\lambda_{bl} a_2} - e^{-\lambda_{bl}(a_1 + a_2 - v)} \quad (8)$$

Let θ be the angular separation between X_1 and X_2. The relationship between the angular separation θ and the correlation coefficient ρ is illustrated in Figure 3 using an example. In the example, the distances from the source transmitter to the two base stations are fixed at $R_1 = 1.2$ and $R_2 = 1.5$ and the base station density is $\lambda_{bs} = 0.3$. In Figure 3a, we fixed the blockage density at $\lambda_{bl} = 0.6$, and the blockage width W is varied. In Figure 3b, $W = 0.5$ and the value of λ_{bl} is varied. Both figures show that ρ decreases with increasing θ. This is because the area v gets smaller as θ increases. As θ approaches 180 degrees, v approaches zero, and the correlation is minimized. The figures show that correlation is more dramatic when W is large, since a single large blockage is likely to simultaneously block both base stations, and when λ_{bl} is small, which corresponds to the case that there are fewer blockages.

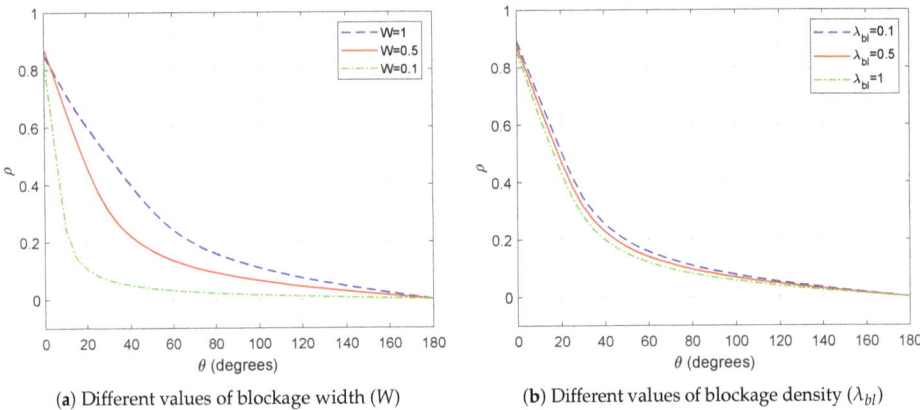

(a) Different values of blockage width (W)   (b) Different values of blockage density (λ_{bl})

Figure 3. The correlation coefficient (ρ) versus the angular separation (θ) between X_1 and X_2.

Figure 4 shows the empirical cumulative distribution function (CDF) of p_{LOS} over 1000 network realizations for first- and second-order macrodiversity, both with and without considering blockage correlation. The distributions are computed by fixing the value of $W = 0.8$ and using two different values of the average number of blockages per base station ($\lambda_{bl}/\lambda_{bs}$). The CDF of p_{LOS} quantifies the likelihood that the p_{LOS} is below some value. The figure shows the probability that p_{LOS} is below some value increases significantly when the number of blockages per base station is high. The effect of correlated blocking is more pronounced when there are fewer blockages per base station. The macrodiversity gain is the improvement in performance for $N = 2$ as compared to $N = 1$, in the figure the macrodiversity gain is higher when the number of blockages per base station is lower even though the amount of reduction in gain due to correlation is higher when $\lambda_{bl}/\lambda_{bs}$ is lower.

Figure 5 shows the variation of p_{LOS} when averaged over 1000 network realizations. In this figure, 1000 p_{LOS} values is found for different 1000 network realization, then the averaged p_{LOS} is calculated for different values of blockage density λ_{bl}. The derivation of the distances for each network realization can be found in Appendix B. The plot shows average p_{LOS} as a function of λ_{bl} while keeping base station density λ_{bs} fixed at 0.3. The spatially averaged p_{LOS} is computed for two different values of blockage width W. Compared to the case of no diversity (when $N = 1$), the second-order macrodiversity can significantly increase p_{LOS}. However, p_{LOS} decreases when blockage size or blockage density is higher. Moreover, correlated blocking reduces the p_{LOS} compared to independent blocking, and larger blockages increase the correlation, since a single large blockage is likely to simultaneously block both base stations. Comparing the two pairs of correlated/uncorrelated blocking curves, the correlation is more dramatic when λ_{bl} is low, since at low λ_{bl} both base stations are typically blocked by the same blockage (located in area v).

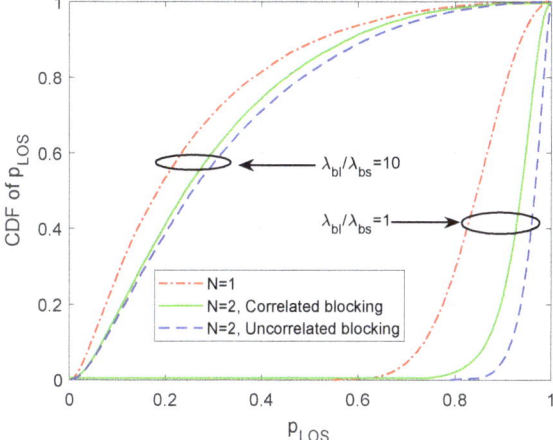

Figure 4. The empirical cumulative distribution function (CDF) of p_{LOS} over 1000 network realizations when $N = 1, 2$, with and without considering blockage correlation at fixed blockage width $W = 0.8$.

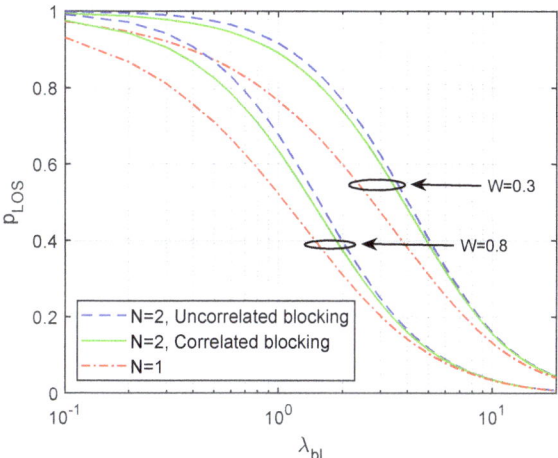

Figure 5. The variation of the spatially averaged p_{LOS} over 1000 network realizations with respect to blockage density λ_{bl} when $N = 1, 2$, with and without considering blockage correlation at fixed base station density $\lambda_{bs} = 0.3$.

4. SNR Distribution

In this section, we consider the distribution of the SNR. Macrodiversity can be achieved by using either diversity combining, where the signals from the multiple base stations are maximum ratio combined, or selection combining, where only the signal with the strongest SNR is used. For nth-order macrodiversity, the SNR with diversity combining is [28]

$$\text{SNR} = \text{SNR}_0 \underbrace{\sum_{i=1}^{n}(1 - B_i)\Omega_i}_{Z} \qquad (9)$$

where $\Omega_i = R_i^{-\alpha}$ is the power gain between the source transmitter Y_0 to the ith base station and SNR_0 is the SNR of an unblocked reference link of unit distance. B_i is used to indicate that the path between Y_0 and X_i is blocked, and thus when $B_i = 1$, Ω_i does not factor into the SNR.

The CDF of SNR, $F_{SNR}(\beta)$, quantifies the likelihood that the combined SNR at the closest n base stations is below some threshold β. If β is interpreted as the minimum acceptable SNR required to achieve reliable communications, then $F_{SNR}(\beta)$ is the *outage probability* of the system $P_o(\beta) = F_{SNR}(\beta)$. The *coverage probability* is the *complimentary* CDF, $P_c(\beta) = 1 - F_{SNR}(\beta)$ and is the likelihood that the SNR is sufficiently high to provide coverage. The *rate* distribution can be found by linking the threshold β to the transmission rate, for instance by using the appropriate expression for channel capacity.

The CDF of SNR evaluated at threshold β is as follows:

$$F_{SNR}(\beta) = P[SNR \le \beta] = P\left[Z \le \frac{\beta}{SNR_0}\right] = F_Z\left(\frac{\beta}{SNR_0}\right). \tag{10}$$

The discrete variable Z represents the sum of the unblocked signals. To find the CDF of Z we need to find the probability of each value of Z, which is found as follows for second-order macrodiversity. The probability that $Z = 0$ can be found by noting that $Z = 0$ when both X_1 and X_2 are blocked. From (1), this is

$$p_Z(0) = p_{B_1,B_2}(1,1) = p_1 p_2 + \rho h. \tag{11}$$

The probability that $Z = \Omega_i, i \in \{1,2\}$ can be found by noting that $Z = \Omega_i$ when only X_i is LOS. From (1), this is

$$p_Z(\Omega_1) = p_{B_1,B_2}(0,1) = q_1 p_2 - \rho h. \tag{12}$$
$$p_Z(\Omega_2) = p_{B_1,B_2}(1,0) = p_1 q_2 - \rho h. \tag{13}$$

Finally, by noting that $Z = \Omega_1 + \Omega_2$ when both X_1 and X_2 are LOS leads to

$$p_Z(\Omega_1 + \Omega_2) = p_{B_1,B_2}(0,0) = q_1 q_2 + \rho h. \tag{14}$$

From (11) to (14), the CDF of Z is found to be:

$$F_Z(z) = \begin{cases} 0 & \text{for } z < 0 \\ p_1 p_2 + \rho h & \text{for } 0 \le z < \Omega_2 \\ p_1 & \text{for } \Omega_2 \le z < \Omega_1 \\ p_1 + q_1 p_2 - \rho h & \text{for } \Omega_1 \le z < \Omega_1 + \Omega_2 \\ 1 & \text{for } z \ge \Omega_1 + \Omega_2. \end{cases} \tag{15}$$

Next, in the case of selection combining, the SNR is [28]

$$SNR = SNR_0 \underbrace{\max\left[(1 - B_1)\Omega_1, (1 - B_2)\Omega_2, \ldots, (1 - B_n)\Omega_n\right]}_{Z} \tag{16}$$

and its CDF, from (11) to (13), is found for second-order macrodiversity to be:

$$F_Z(z) = \begin{cases} 0 & \text{for } z < 0 \\ p_1 p_2 + \rho h & \text{for } 0 \le z < \Omega_2 \\ p_1 & \text{for } \Omega_2 \le z < \Omega_1 \\ 1 & \text{for } z \ge \Omega_1. \end{cases} \tag{17}$$

Figure 6 is an example showing the effect that the value of the correlation coefficient ρ has upon the CDF of SNR. The curves were computed by placing the base stations at distances $R_1 = 2$ and $R_2 = 5$, and fixing the values of $\alpha = 2$ and $SNR_0 = 15$ dB. The values of q_i and p_i were computed using (3) and (4) respectively, by assuming $W = 0.6$, $\lambda_{bl} = 0.3$. The CDF is found assuming values of ρ between $\rho = 0$ to $\rho = 0.8$ in increments of 0.1; the value of ρ can be adjusted by varying the angle θ between the two base stations. The dashed red line represents the case that $\rho = 0$, corresponding to uncorrelated blocking. The solid blue lines correspond to positive values of ρ in increments of 0.1, where the thinnest line corresponds to $\rho = 0.1$ and the thickest line corresponds to $\rho = 0.8$.

Figure 6 shows a first step up at 9.7 dB, and the increment of the step is equal to the probability that both base stations are NLOS. The magnitude of the step gets larger as the blocking is more correlated, because correlation increases the chance that both base stations are NLOS (i.e., $p_{B_1,B_2}(1,1)$). The next step up occurs at 12.7 dB, which is the SNR when just one of the two closest base stations is blocked, which in this case is the closest base station X_1. The next step at 14.5 dB represents the case when only X_2 is blocked, The magnitude of the two jumps is equal to the probability that only the corresponding one base station is LOS, and this magnitude decreases with positive correlation, because if one base station is LOS the other one is NLOS. Finally, there is a step at 15.2 dB, which corresponds to the case that both base stations are LOS. Notice that when $\rho = 0.8$, the two middle steps merge. This is because for such a high value of, it is impossible for just one base station to be blocked, and most likely that both base stations are blocked, so the curve goes directly from SNR = 9.7 dB to SNR = 15.2 dB.

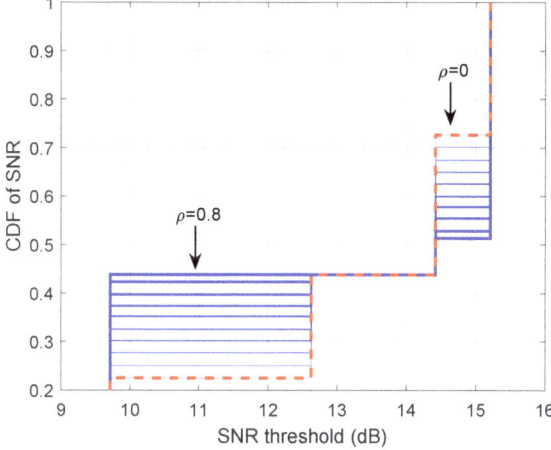

Figure 6. The CDF of the Signal to Noise Ratio (SNR) $F_{SINR}(\beta)$ using diversity combining for fixed location of X_1 and X_2 for different values of ρ. The dashed red line shows the CDF when $\rho = 0$, and the solid blue lines correspond to positive values of ρ in increments of 0.1.

Figure 7 shows the CDF of SNR over 1000 network realizations for diversity combining and two different values of W when $\lambda_{bs} = 0.4$ and $\lambda_{bl} = 0.6$. In addition, SNR_0 and the path loss α are fixed at 15 dB and 3 respectively for the remaining figures in this paper. It can be observed that the CDF increases when blockage size is larger. Compared to the case when $N = 1$, the use of second-order macrodiversity decreases the SNR distribution. When compared to uncorrelated blocking, correlation decreases the gain of macrodiversity for certain regions of the plot, particularly at low values of SNR threshold, corresponding to the case when both base stations are blocked. Similar to p_{LOS}, the correlation increases with blockage size. However, the macrodiversity gain is slightly higher when blockage width W is smaller.

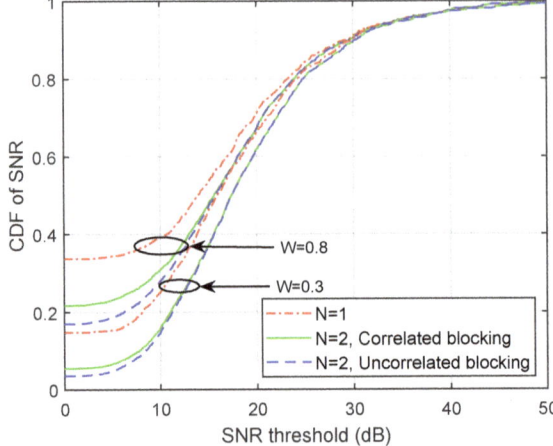

Figure 7. The distribution of SNR over 1000 network realizations when $N = 1, 2$ using diversity combining, with and without considering blockage correlation at fixed values of blockage density $\lambda_{bl} = 0.6$ and base station density $\lambda_{bs} = 0.3$.

Figure 8 shows the effect of combining scheme and λ_{bl} on SNR outage probability at threshold $\beta = 10$ dB. As shown in the figure, the outage probability increases when λ_{bl} increases in all of the given scenarios. When $\lambda_{bl} = 0$, first- and second-order selection combining perform identically. This is because X_1 is never blocked. However, as λ_{bl} increases, the gain of both selection combining and diversity combining increase. At high λ_{bl} the combining scheme is less important, in which case the paths to X_1 and X_2 are always blocked regardless of the chosen combining scheme. The reduction in gain due to correlation is slightly higher when using selection combining. From Equation (17) this is because the step when both base stations are blocked is wider compared to diversity combining case.

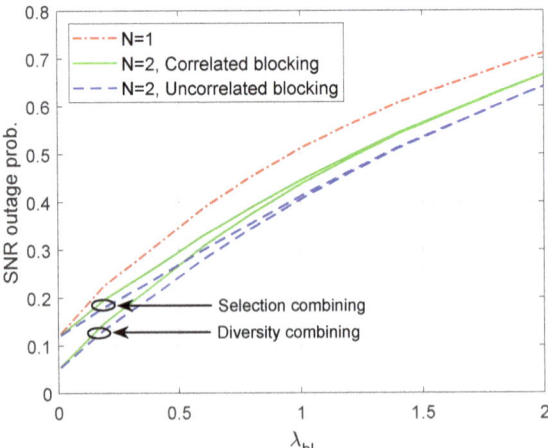

Figure 8. The SNR outage probability at threshold $\beta = 10$ dB with respect to λ_{bl} when $N = 1, 2$, with and without considering blockage correlation at fixed values of blockage density $\lambda_{bs} = 0.3$ and blockage width $W = 0.8$.

5. Real Data Validation

To validate our framework, we consider a region of West Virginia University campus as shown in Figure 9 with base stations locations drawn from a PPP and a randomly placed user. The exterior walls of the buildings highlighted in red color are considered to be the blockages. The equivalent parameters for the statistical analysis introduced by this paper are obtained by calculating the number of buildings, the area of each building, and the total area of the region. The average blockage width (W) is found from the areas of the individual buildings (A_i), such that the width of each blockage $W_i = 2\sqrt{A_i/\pi}$, while the blockage density is found as the the number of buildings divided by the total region area.

Figure 9. Map of West Virginia University (WVU) downtown campus. The red-highlighted buildings are the blockages, and the base stations and user are randomly placed over the region.

Figure 10 shows the empirical CDF of SNR over 1000 network realizations computed using our statistical analysis and computed using the actual data. The total region area is found to be 335,720 m², the number of buildings is 49, the average building width is $W = 33$ m, λ_{bl} is the ratio of number of buildings to the total area, and $\lambda_{bs} = 3\lambda_{bl}$. We limited the environment to be outdoor by allowing the base stations and user to only be located outside buildings. It can be observed that the analysis approximates the performance in the real scenario very well. Compared to the curves representing the analysis when $N = 2$, it is clear that the real data model when $N = 2$ is closer to the case when considering correlated blocking compared to the case assuming independent blocking. This is because one building can simultaneously block more than one base station. In the actual region, the blockages have different sizes and orientations, this is in contrast with our model, which assumes a constant blockage size and orientation. Due to these differences, there is a small different between the statistical model and the real data based model as shown in the figure.

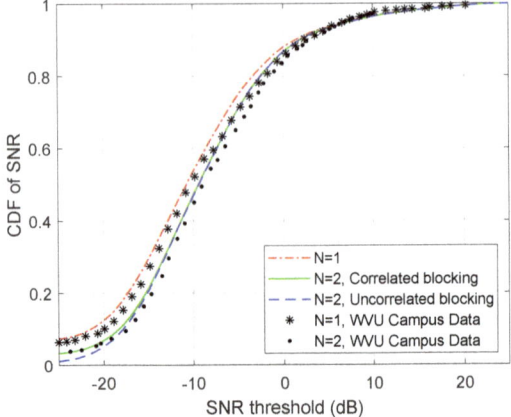

Figure 10. The distribution of SNR over 1000 network realizations when $N = 1, 2$ using diversity combining, plotted using the real data model and the analytical model.

6. SINR Outage Analysis

Thus far, we have not assumed any interfering transmitters in the system. In practice, the received signal is also affected by the sum interference. The goal of this section is to formulate the CDF of SINR for second-order macrodiversity. SINR for first-order macrodeiversity along with blockage correlation between interferers has been considered in [15]. In this section, we assume each neighboring cell has a single interfering mobile, which is located uniformly within a disk of radius r around the base station. Assuming a perfect packing of cells, $r = (\lambda_{bs}\pi)^{-1/2}$, which is the average cell radius. We explicitly consider the interference from the M closest neighboring cells. The interference from more distant cells is considered to be part of the thermal noise. Let Y_j for $j = 1, 2, .., M$ indicate the interfering transmitters and their locations. Recall that $j = 0$ indicates the source transmitter Y_0. The distance from the jth transmitter to the ith base station is denoted by $R_{i,j}$.

To calculate SINR and its distribution, we first define a matrix **B** which indicates the blocking state of the paths from Y_j for $j = 0, 2, .., M$ to X_i for $i = 1, 2$. **B** is a Bernoulli Matrix of size 2 by $(M + 1)$ elements. Each column in **B** contain elements $B_{1,j}$ and $B_{2,j}$ which indicate the blocking states of the paths from Y_j to X_1 and X_2 respectively; i.e, the first column in **B** contains the pair of Bernoulli random variables $B_{1,0}$ and $B_{2,0}$ that indicates the blocking state of the paths from Y_0 to X_i for $i = 1, 2$. There are $(M + 1)$ pairs of Bernoulli random variables, and each pair is correlated with correlation coefficient ρ_j. Because the $2(M + 1)$ elements of **B** are binary, there are $2^{2(M+1)}$ possible combinations of **B**. However, it is possible for different realizations of **B** to correspond to the same value of SINR. For example, when X_1 and X_2 are both blocked from Y_0, the SINR will be the same value regardless of the blocking states of the interfering transmitters. Define $\mathbf{B}^{(n)}$ for $n = 1, 2, ..., 2^{2(M+1)}$ to be the nth such combination of **B**. Similar to Section 3, let $p_{B_{1,j}, B_{2,j}}(b_{1,j}^{(n)}, b_{2,j}^{(n)})$ be the joint probability of $B_{1,j}$ and $B_{2,j}$ which are the elements of the jth column of $\mathbf{B}^{(n)}$. The probability of $\mathbf{B}^{(n)}$ is given by

$$P(\mathbf{B}^{(n)}) = \prod_{j=0}^{M} p_{B_{1,j}, B_{2,j}}(b_{1,j}^{(n)}, b_{2,j}^{(n)}) \tag{18}$$

The SINR of a given realization $\mathbf{B}^{(n)}$ at base station X_i is given by

$$\text{SINR}_i^{(n)} = \frac{(1 - B_{i,0}^{(n)})\Omega_{i,0}}{\text{SNR}_0^{-1} + \sum_{j=1}^{M}(1 - B_{i,j}^{(n)})\Omega_{i,j}} \tag{19}$$

where $\Omega_{i,j} = R_{i,j}^{-\alpha}$ is the path gain from the jth transmitter at the ith base station. The SINR of the combined signal considering selective combining is expressed as

$$\text{SINR}^{(n)} = \max\left(\text{SINR}_1^{(n)}, \text{SINR}_2^{(n)}\right) \tag{20}$$

When considering diversity combining (20) changes to

$$\text{SINR}^{(n)} \leq \text{SINR}_1^{(n)} + \text{SINR}_2^{(n)} \tag{21}$$

As described in [29], correlated interference tends to make the combined SINR less than the sum of the individual SINRs. The bound in (21) is satisfied with equality when the interference is independent at the two base stations.

To generalize the formula for any realization, there is a particular $\text{SINR}^{(n)}$ associated with each $\mathbf{B}^{(n)}$. However, as referenced above, multiple realizations of $\mathbf{B}^{(n)}$ may result in the same SINR. Let $\text{SINR}^{(k)}$ be the kth realization of SINR. Its probability is

$$P\left(\text{SINR}^{(k)}\right) = \sum_{n:\text{SINR}=\text{SINR}^{(k)}} P\left(\mathbf{B}^{(n)}\right) \tag{22}$$

Figure 11 shows the distributions of SINR for $M = 5$ and $M = 0$ (which is SNR) at fixed values of $\lambda_{bs} = 0.3$, $\lambda_{bl} = 0.6$, and $W = 0.6$. The distributions are computed for first- and second-order macrodiversity. It can be observed that macrodiversity gain is reduced when interference is considered. This is because of the increase in sum interference due to macrodiversity, which implies that p_{LOS} alone as in [13] may not be sufficient to predict the performance of the system especially when there are many interfering transmitters. Study of higher order macrodiversity to identify the minimum order of macrodiversity to achieve a desired level of performance in the presence of interference is left for future work.

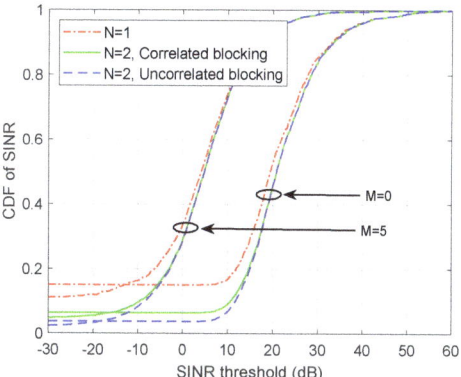

Figure 11. The distribution of Signal to Interference and Noise Ratio (SINR) over 1000 network realizations using diversity combining for different values of number of interfering transmitters. The curves are computed when $N = 1, 2$, with and without considering blockage correlation, at fixed values of $\lambda_{bs} = 0.3$, $\lambda_{bl} = 0.6$, and $W = 0.6$.

Figure 12 shows the variation of SINR outage probability with respect to the number of interfering transmitters M. The curves are computed for low and high values of λ_{bl}, while keeping λ_{bs} and W fixed at 0.8 and 0.6 respectively. It can be seen that the outage probability increases when M increases. Due to the fact that interference tends to also be blocked, unlike SNR and p_{LOS}, increasing the λ_{bl} decreases the outage probability. Similar to Figure 11, the macrodiversity gain decreases significantly when M increases. It can be seen that $N = 2$ curves reaches the case when $N = 1$ for $M = 6$. Compared to uncorrelated blocking, the curves considering correlated blocking matches the uncorrelated cases for high value of M, since the interfering transmitters are placed farther than source transmitter and their overlapping area is less dominant.

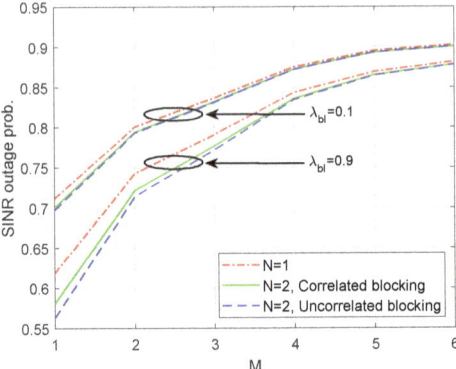

Figure 12. The outage probability of SINR at threshold $\beta = 15$ dB versus the number of interfering transmitters (M), when $N = 1, 2$, with and without considering blockage correlation, at fixed values of $\lambda_{bs} = 0.8$ and $W = 0.6$.

7. Conclusions

We have proposed a framework to analyze the second-order macrodiversity gain for an mmWave cellular system in the presence of correlated blocking. Correlation is an important consideration for macrodiversity because a single blockage can block multiple base stations, especially if the blockage is sufficiently large and the base stations sufficiently close. The assumption of independent blocking leads to an incorrect evaluation of macrodiversity gain of the system. By using the methodology in this paper, the correlation between two base stations is found and factored into the analysis. The paper considered the distributions of LOS probability, SNR, and, when there is interference, the SINR. The framework was confirmed by comparing the analysis to a real data model. We show that correlated blocking decreases the macrodiversity gain. We also study the impact of blockage size and blockage density. We show that blockage can be both a blessing and a curse. On the one hand, the signal from the source transmitter could be blocked, and on the other hand, interfering signals tend to also be blocked, which leads to a completely different effect on macrodiversity gains.

The analysis can be extended in a variety of ways. In Section 6, we have already shown that any number of interfering transmitters can be taken in to account. While this paper has focused on the extreme case that LOS signals are AWGN while NLOS signals are completely blocked, it is possible to adapt the analysis to more sophisticated channels, such as those where both LOS and NLOS signals are subject to fading and path loss, but the fading and path loss parameters are different depending on the blocking state. See, for instance, [17] for more detail. We may also consider the use of directional antennas, which will control the effect of interference [30].

Finally, while this paper focused on second-order macrodiversity, the study can be extended to the more general case of an arbitrary macrodiversity order. Such a study could identify the minimum macrodiversity order required to achieve desired performance in the presence of interference. We anticipate that when more than two base stations are connected, the effects of correlation on macrodiversity gain will increase and the effect of interference will decrease. This is because the likelihood that two base stations are close together increases with the number of base stations and the ratio of the number of connected base stations to the number of interfering transmitters will increase.

Author Contributions: Conceptualization, E.H. and M.C.V.; methodology, E.H. and M.C.V.; software, E.H.; validation, E.H.; formal analysis, E.H. and M.C.V.; investigation, E.H.; resources, M.C.V.; data curation, E.H.;

writing–original draft preparation, E.H.; writing–review and editing, M.C.V.; visualization, E.H.; supervision, M.C.V.; project administration, M.C.V.; funding acquisition, M.C.V.

Funding: This research was supported in part by the National Science Foundation under Grant 1650474.

Conflicts of Interest: The authors declare no conflict of interest. The funders had no role in the design of the study; in the collection, analyses, or interpretation of data; in the writing of the manuscript, or in the decision to publish the results.

Appendix A

As in [27], the correlation coefficient between B_1 and B_2 is given by

$$\rho = \frac{E[B_1 B_2] - E[B_1]E[B_2]}{\sqrt{\sigma_{B_1}^2 \sigma_{B_2}^2}} \tag{A1}$$

where the expected value and the variance of the Bernoulli variable B_i is given by [27]

$$E[B_i] = p_i \tag{A2}$$
$$\sigma_{B_i}^2 = p_i q_i. \tag{A3}$$

By substituting (A2) and (A3) into (A1) and solving for $E[B_1 B_2]$,

$$E[B_1 B_2] = p_1 p_2 + \rho \sqrt{p_1 p_2 q_1 q_2} = p_1 p_2 + \rho h. \tag{A4}$$

As in [27], we can relate $p_{B_1,B_2}(b_1,b_2)$ to $E[B_1 B_2]$ as follows:

$$E[B_1 B_2] = \sum_{b_1}\sum_{b_2} b_1 b_2 p_{B_1,B_2}(b_1,b_2) = p_{B_1,B_2}(1,1), \tag{A5}$$

where solving the sum relies there being only one nonzero value for $b_1 b_2$. By solving for $p_{B_1,B_2}(1,1)$ and using (A4),

$$p_{B_1,B_2}(1,1) = p_1 p_2 + \rho h. \tag{A6}$$

We can relate $p_{B_1,B_2}(b_1,b_2)$ to $E[B_1]$ as follows:

$$E[B_1] = \sum_{b_1}\sum_{b_2} b_1 p_{B_1,B_2}(b_1,b_2)$$
$$= p_{B_1,B_2}(1,1) + p_{B_1,B_2}(1,0). \tag{A7}$$

Solving for $p_{B_1,B_2}(1,0)$,

$$p_{B_1,B_2}(1,0) = E[B_1] - p_{B_1,B_2}(1,1) = p_1 q_2 - \rho h. \tag{A8}$$

Similarly, it can be shown that

$$p_{B_1,B_2}(0,1) = q_1 p_2 - \rho h. \tag{A9}$$

Finally, since [27]

$$\sum_{b_1}\sum_{b_2} p_{B_1,B_2}(b_1,b_2) = 1, \tag{A10}$$

it follows that

$$p_{B_1,B_2}(0,0) = 1 - p_{B_1,B_2}(1,0) - p_{B_1,B_2}(0,1) - p_{B_1,B_2}(1,1)$$
$$= q_1 q_2 + \rho h. \tag{A11}$$

Appendix B

As in [13], the pdf of the smallest distance R_1 is

$$f(r_1) = 2\pi \lambda r_1 e^{-\lambda \pi r_1^2} \tag{A12}$$

for $r_1 \geq 0$. From (A12), we can derive the conditional CDF of R_i given R_{i-1} as

$$F_{R_i}(r_i | R_{i-1} = r_{i-1}) = 1 - e^{\lambda \pi (r_i^2 - r_{i-1}^2)} \tag{A13}$$

To generate random variables $r_1, ..., r_N$, let $x_i \sim U(0,1)$,

$$x_i = F_{R_i}(r_i | R_{i-1} = r_{i-1}) = 1 - e^{\lambda \pi (r_i^2 - r_{i-1}^2)} \tag{A14}$$

Solving for r_i,

$$r_i = \sqrt{-\frac{1}{\lambda \pi} \ln(1 - x_i) + r_{i-1}^2} \tag{A15}$$

where $r_0 = 0$. Start by generating x_i as uniform random variables, then recursively substitute each one in (A15) to get the desired random variable r_i.

References

1. Rappaport, T.S.; Sun, S.; Mayzus, R.; Zhao, H.; Azar, Y.; Wang, K.; Wong, G.N.; Schulz, J.K.; Samimi, M.; Gutierrez, F. Millimeter wave mobile communications for 5G cellular: It will work. *IEEE Access* **2013**, *1*, 335–349. [CrossRef]
2. Akdeniz, M.R.; Liu, Y.; Samimi, M.K.; Sun, S.; Rangan, S.; Rappaport, T.S.; Erkip, E. Millimeter Wave Channel Modeling and Cellular Capacity Evaluation. *IEEE J. Sel. Areas Commun.* **2014**, *32*, 1164–1179. [CrossRef]
3. Andrews, J.G.; Bai, T.; Kulkarni, M.N.; Alkhateeb, A.; Gupta, A.K.; Heath, R.W., Jr. Modeling and Analyzing Millimeter Wave Cellular Systems. *IEEE Trans. Commun.* **2017**, *65*, 403–430. [CrossRef]
4. Park, J.; Andrews, J.G.; Heath, R.W. Inter-Operator Base Station Coordination in Spectrum-Shared Millimeter Wave Cellular Networks. *IEEE Trans. Cogn. Commun. Netw.* **2018**, *4*, 513–528. [CrossRef]
5. Singh, S.; Kulkarni, M.N.; Ghosh, A.; Andrews, J.G. Tractable Model for Rate in Self-Backhauled Millimeter Wave Cellular Networks. *IEEE J. Sel. Areas Commun.* **2015**, *33*, 2196–2211. [CrossRef]
6. Jurdi, R.; Gupta, A.K.; Andrews, J.G.; Heath, R.W. Modeling Infrastructure Sharing in mmWave Networks With Shared Spectrum Licenses. *IEEE Trans. Cogn. Commun. Netw.* **2018**, *4*, 328–343. [CrossRef]
7. Rappaport, T.; Heath, R.W., Jr.; Daniels, R.C.; Murdock, J.N. *Millimeter Wave Wireless Communications*; Pearson Education: Upper Saddle River, NJ, USA, 2014.
8. Fisher, R. 60 GHz WPAN Standardization within IEEE 802.15.3c. In Proceedings of the International Symposium on Signals, Systems and Electronics, Montreal, QC, Canada, 30 July–2 August 2007; pp. 103–105.
9. Singh, S.; Ziliotto, F.; Madhow, U.; Belding, E.; Rodwell, M. Blockage and directivity in 60 GHz wireless personal area networks: From cross-layer model to multihop MAC design. *IEEE J. Sel. Areas Commun.* **2009**, *27*, 1400–1413. [CrossRef]
10. Niu, Y.; Li, Y.; Jin, D.; Su, L.; Wu, D. Blockage Robust and Efficient Scheduling for Directional mmWave WPANs. *IEEE Trans. Veh. Technol.* **2015**, *64*, 728–742. [CrossRef]
11. Bai, T.; Vaze, R.; Heath, R.W., Jr. Analysis of Blockage Effects on Urban Cellular Networks. *IEEE Trans. Wirel. Commun.* **2014**, *13*, 5070–5083. [CrossRef]

12. Choi, J. On the Macro Diversity With Multiple BSs to Mitigate Blockage in Millimeter-Wave Communications. *IEEE Commun. Lett.* **2014**, *18*, 1653–1656. [CrossRef]
13. Gupta, A.K.; Andrews, J.G.; Heath, R.W. Macrodiversity in Cellular Networks With Random Blockages. *IEEE Trans. Wirel. Commun.* **2018**, *17*, 996–1010. [CrossRef]
14. Gupta, A.K.; Andrews, J.G.; Heath, R.W. Impact of Correlation between Link Blockages on Macro-Diversity Gains in mmWave Networks. In Proceedings of the IEEE International Conference on Communications Workshops, Kansas City, MO, USA, 20–24 May 2018.
15. Hriba, E.; Valenti, M.C. The Potential Gains of Macrodiversity in mmWave Cellular Networks with Correlated Blocking. In Proceedings of the IEEE Military Communications Conference (MILCOM), Norfolk, VA, USA, 12–14 November 2019.
16. Venugopal, K.; Valenti, M.C.; Heath, R.W., Jr. Device-to-Device Millimeter Wave Communications: Interference, Coverage, Rate, and Finite Topologies. *IEEE Trans. Wirel. Commun.* **2016**, *15*, 6175–6188. [CrossRef]
17. Hriba, E.; Valenti, M.C.; Venugopal, K.; Heath, R.W. Accurately Accounting for Random Blockage in Device-to-Device mmWave Networks. In Proceedings of the IEEE Global Telecommunications Conference (GLOBECOM), Singapore, 4–8 December 2017.
18. Hriba, E.; Valenti, M.C. The Impact of Correlated Blocking on Millimeter-Wave Personal Networks. In Proceedings of the IEEE Military Communications Conference (MILCOM), Los Angeles, CA, USA, 29–31 October 2018.
19. Thornburg, A.; Heath, R.W. Ergodic Rate of Millimeter Wave Ad Hoc Networks. *IEEE Trans. Wirel. Commun.* **2018**, *17*, 914–926. [CrossRef]
20. Baccelli, F.; Błaszczyszyn, B.; Muhlethaler, P. An Aloha protocol for multihop mobile wireless networks. *IEEE Trans. Inf. Theory* **2006**, *52*, 421–436. [CrossRef]
21. Venugopal, K.; Valenti, M.C.; Heath, R.W., Jr. Interference in finite-sized highly dense millimeter wave networks. In Proceedings of the Information Theory and Applications Workshop (ITA), San Diego, CA, USA, 1–6 February 2015, pp. 175–180.
22. Aditya, S.; Dhillon, H.S.; Molisch, A.F.; Behairy, H. Asymptotic Blind-Spot Analysis of Localization Networks Under Correlated Blocking Using a Poisson Line Process. *IEEE Wirel. Commun. Lett.* **2017**, *6*, 654–657. [CrossRef]
23. Aditya, S.; Dhillon, H.S.; Molisch, A.F.; Behairy, H.M. A Tractable Analysis of the Blind Spot Probability in Localization Networks Under Correlated Blocking. *IEEE Trans. Wirel. Commun.* **2018**, *17*, 8150–8164. [CrossRef]
24. Talarico, S.; Valenti, M.C.; Di Renzo, M. Outage Correlation in Finite and Clustered Wireless Networks. In Proceedings of the IEEE International Symposium on Personal, Indoor and Mobile Radio Communications (PIMRC), Bologna, Italy, 9–12 September 2018; pp. 1–7.
25. Samuylov, A.; Gapeyenko, M.; Moltchanov, D.; Gerasimenko, M.; Singh, S.; Himayat, N.; Andreev, S.; Koucheryavy, Y. Characterizing Spatial Correlation of Blockage Statistics in Urban mmWave Systems. In Proceedings of the IEEE Globecom Workshops, Washington, DC, USA, 4–8 December 2016; pp. 1–7.
26. Snyder, D.L.; Miller, M.I. *Random Point Processes in Time and Space*, 2nd ed.; Springer Texts in Electrical Engineering: New York, NY, USA, 1991.
27. Leon-Garcia, A. *Probability and Random Processes For Electrical Engineering*, 3rd ed.; Pearson: Boston, MA, USA, 2008.
28. Goldsmith, A. *Wireless Communications*; Cambridge University Press: New York, NY, USA, 2005.
29. Tanbourgi, R.; Dhillon, H.S.; Andrews, J.G.; Jondral, F.K. Effect of Spatial Interference Correlation on the Performance of Maximum Ratio Combining. *IEEE Trans. Wirel. Commun.* **2014**, *13*, 3307–3316. [CrossRef]
30. Torrieri, D.; Talarico, S.; Valenti, M.C. Analysis of a Frequency-Hopping Millimeter-Wave Cellular Uplink. *IEEE Trans. Wirel. Commun.* **2016**, *15*, 7089–7098. [CrossRef]

© 2019 by the authors. Licensee MDPI, Basel, Switzerland. This article is an open access article distributed under the terms and conditions of the Creative Commons Attribution (CC BY) license (http://creativecommons.org/licenses/by/4.0/).

Article

A Software-Defined Radio for Future Wireless Communication Systems at 60 GHz

Luis Duarte [1,2,3], Rodolfo Gomes [4], Carlos Ribeiro [1,2] and Rafael F. S. Caldeirinha [1,2,*]

1 Instituto de Telecomunicações, 2411-901 Leiria, Portugal; luis.duarte@ipleiria.pt (L.D.); carlos.ribeiro@ipleiria.pt (C.R.)
2 Polytechnic Institute of Leiria, 2411-901 Leiria, Portugal
3 Department of Eletronics, Telecommunications and Informatics, University of Aveiro, 3810-193 Aveiro, Portugal
4 TWEVO Lda., 3030-199 Coimbra, Portugal; rgomes@twevo.net
* Correspondence: rafael.caldeirinha@ipleiria.pt

Received: 6 November 2019; Accepted: 3 December 2019; Published: 6 December 2019

Abstract: This paper reports on a complete end-to-end 5G mmWave testbed fully reconfigurable based on a FPGA architecture. The proposed system is composed of a baseband/low-IF processing unit, and a mmWave RF front-end at both TX/RX ends. In particular, the baseband unit design is based on a typical agile digital IF architecture, enabling on-the-fly modulations up to 256-QAM. The real-time 5G mmWave testbed, herein presented, adopts OFDM as the transmission scheme waveform, which was assessed OTA by considering the key performance indicators, namely EVM and BER. A detailed overview of system architecture is addressed together with the hardware considerations taken into account for the mmWave testbed development. Following this, it is demonstrated that the proposed testbed enables real-time multi-stream transmissions of UHD video content captured by nine individual cameras, which is in fact one of the killing applications for 5G.

Keywords: mmWave; 5G; OFDM; SDR; testbed; FPGA; RF frontends; UHD stream

---

1. Introduction

Millimeter wave (mmWave) communications are envisaged to be integrated in the upcoming generation of mobile networks, namely eMBB, one of the 5G verticals. This has been seen as the solution to increase the overall capacity of mobile radio cells, enabling the support of multi-Gigabit/s transmissions towards mobile equipments. Due to the large available bandwidth (55–66 GHz), mmWave band is, thus, very attractive for future 5G wireless communication systems, which might provide transmission data rates over 10 Gbps and network latency below 1 ms. However, even if a 2 GHz channel in the 60 GHz band is used to transmit data employing both 4 and 16-Quadrature Amplitude Modulation (QAM) modulations, data rate would still be limited to 4 and 8 Gbps, respectively. Therefore, there is demand for improving both system reliability and data rates. In this context, it is introduced in this work a novel software-define radio (SDR) mmWave testbed, aimed to tackle the 5G communication requirements outlined in [1].

With mmWave testbeds importance in mind, work at 60 GHz can be found in the literature [2–4]. However, such works were conducted with universal software radio peripheral (USRP) that offer limited bandwidths of around 25 MHz capabilities, which do not fulfill the 5G foreseen multi-Gigabits scenarios. To this extent, this work discusses the implementation of a real-time software-defined radio mmWave testbed that can cope with the next wireless generation needs.

A software-defined radio (SDR) is a common term given to a system which employs the majority of physical layer functionalities using digital signal processing algorithms implemented in an embedded system with the aid of a specific software. In this context, typically, analog stages, such as mixing, amplification, filtering, modulators/demodulators, and essential to establish a wireless radio link, are digitally implemented rather than using discrete analog hardware components. Therefore, an ideal SDR is composed of very reduced hardware at the RF front-end, i.e., only an antenna and a very high speed sampler that is capable of capturing and digitizing a wideband radio signals [5]. However, relatively large coverage distances might only be attained by the employment of amplifiers prior to both DAC/ADC stages, as depicted in Figure 1. For example, at the receiver, a Low-Noise Amplifier (LNA) must be considered to reduce the converter quantization noise and thus maximize the Signal-to-Noise Ratio (SNR) at the digital domain (after sampling); only then is the signal converted to its binary form. Next, data are processed on a number of dedicated computational units, inside the embedded system, enabling the implementation of crucial methods for demodulation, synchronization, and decoding, which are required to recover the transmitted information from advanced modulation techniques.

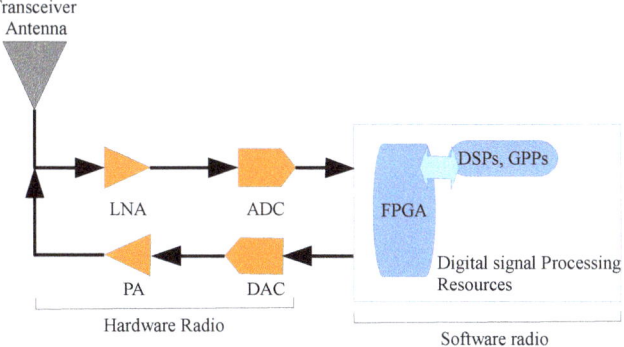

Figure 1. Block diagram of the ideal software-defined radio transceiver (image adapted from [6]).

Nevertheless, such architecture inherits the precision limitation from the digital domain representation, as well as it adds computation complexity to such digital domain. On the other hand, translating hardware discrete analog components functionalities into an embedded platform brings several advantages for a wireless communication system designer [7]:

1. **Design Flexibility:** With the recent availability of very fast and accurate DACs/ADCs, traditional hardware circuitry such as filtering and analog modulators/demodulators can now be tuned.
2. **Reliability:** In general, hardware component functionalities vary according to temperature, manufacturing variations, and ageing, which might lead to performance losses for the communication system. On the other hand, if such functionalities are implemented in a SDR platform, system's performance is expected to remain the same over the years.
3. **Upgradability:** Upgrading and improving a wireless system that is composed of only discrete components usually means replacing hardware, whereas using SDR a particular system might be upgraded without any additional cost, which lead to a more cost effective solution.
4. **Re-usability:** Once a PHY is designed and tested in software, its deployment can be implemented in several hardware platforms.
5. **Reconfigurability:** Current SDR implementations offers the capability of reconfigurable hardware to support several wireless standards and waveforms.
6. **Lower Cost:** Fewer hardware components are necessary to be acquired in order to build a communication system from the scratch. Such devices are usually more expensive than the acquisition of SDR products.

A complete description of the developed SDR is outlined in Section 2, where special focus is given to the adopted firmware connectivity between the embedded system and both Giga Samples Per Second (GSPS) DAC/ADC boards. In addition, an extensive SDR characterization in terms of SNR, Spurious-Free Dynamic Range (SFDR), Signal-to-Noise and Distortion Ratio (SINAD), and Effective Number Of Bits (ENOB) is given, along with the proposed hardware design for Orthogonal Frequency Division Multiplexing (OFDM) systems based on the LTE-Advanced downlink PHY layer is given in Section 3. The OFDM performance in both Back-to-Back (B2B) and Over-The-Air (OTA) at 60 GHz scenarios was assessed in terms of Error Vector Magnitude (EVM) and the received scatter constellation plots for 4-/256-QAM. Finally, a showcase considering the proposed mmWave testbed with Ultra-High-Definition (UHD) multi-stream video, as usage case application, is outlined in Section 4.

2. 5G Testbed Overview

The proposed system was inspired on the previous Figure 1 and is composed of a baseband/low-Intermediate Frequency (IF) processing unit and a RF front-end at both TX/RX ends, as illustrated in Figure 2. In fact, it was the result of SDR OFDM integration with a RF front-end designed for mmWave. In Figure 2, it is evident that the low-IF waveform generation/demodulation is completely independent from the RF up-conversion/down-conversion. This modular approach was adopted to enhance the flexibility of the prototype radio system to enable signal transmissions at distinct frequency bands. That is, the mmWave carrier frequency can be tuned (to match other RF communication systems standards) by only changing the RF front-end.

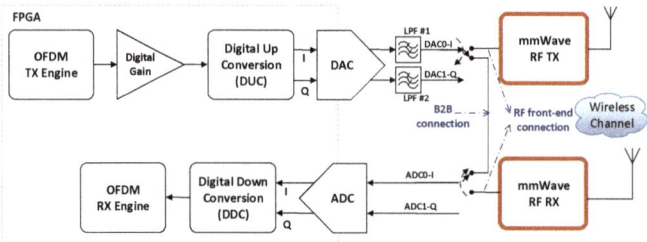

Figure 2. Block diagram of the proposed 5G mmWave testbed.

This SDR system (Figure 2) is composed of an embedded platform engine called Field Programmable Gate Array (FPGA) [8], and at least by one pair DAC/ADC responsible for signal translation between the digital and analog domains. A FPGA is an integrated circuit intended for general purpose usage, which contains programmable logic blocks that form higher-level functions (multiplexers, adders, multipliers and RAM memories) and a reconfigurable interconnection between them [9].

FPGA platforms bring three main benefits for the designer: faster time-to-market (no layout for manufacturing needed), simpler design (the software handles much of the routing, placement, and timing) and re-programmability. Traditionally, all the above features would become available at the expense of chip performance, e.g., clock speed. However, in recent *Xilinx's* FPGA technology digital clock values can now go over the 500 MHz performance barrier [9].

Consequently, FPGA technology was used for the 5G mmWave wireless communication that we propose especially due to their faster time-to-market and reduced cost characteristics. Due to its primal importance in this work, a brief overview of FPGA considerations is given in Section 2.1.

2.1. FPGA

One of the main challenges faced by embedded system developers is the diversity of external I/O interfacing requirements, which might be Ethernet, optical, analog conversion or Gigabit serial [10]. Thus, to enable data transfer between the embedded system and other hardware devices, the creation of external I/O interface is needed. Nowadays, FPGA come with one or more expansion slots for daughter boards (mezzanine module), which can increase the I/O possibilities. Such expansion slots allow FPGA designers to not rely solely on each FPGA I/Os and decrease system development costs. Without it, designers would need to change FPGA development board whenever system enhancements require additional interfaces [11].

To overcome this, the VITA 57 FPGA Mezzanine Card (FMC) connectivity ANSI/VITA standard was developed in 2008, to provide a modular standard mezzanine I/O interfacing solution to a FPGA located on a carrier board [11]. This brought several benefits such as maximum data throughput of 40 Gbps, low latency, reduced design complexity, and reduction of system cost [10].

Such standard foresees two types of expansion slots, also called FMC, which are usually populated on FPGA commercial solutions (LPC and HPC). While the first FMC solution has 160 pins, the second one has 400 available pins. Moreover, the Low Pin Count (LPC) connector provides 68/34 user-defined signals for single-ended and differential signals, respectively, as well as one serial transceiver pair, clocks, Joint Test Access Group (JTAG), and Inter-Integrated Circuit (I2C) interfaces. The High Pin Count (HPC) maintains precedent mentioned functionalities while providing twice the user-defined signals of the LPC connector also with 10 serial transceiver pairs.

At the time of this work, the most recent and powerful *Xilinx* FPGA (with more logic resources) available on the market is the Virtex FPGA's family, namely Ultrascale [9]. For example, the XCVU9P-L2FLGA2104E chip provides more than 2000 k logic cells [12]. However, it is relatively expensive, and thus it was decided to acquire the Virtex-7 VC707 evaluation kit instead [13] (see Figure 3). This board is characterized as highly-flexible, having a high-speed serial XC7VX485T-2FFG1761C chip-set, and a good compromise between cost and available logical resources.

Figure 3. VC707 FPGA with both FMC HPC used by ADC/DAC daughterboards.

2.2. ADC/DAC

The selection of both DAC and ADC was based on the criteria of speed, bit resolution together with VC707 FPGA board compatibility. According to the *Xilinx* documentation, the FMC 230 (DAC) [14] and FMC 126 (ADC) [15] from 4DSP are the only converters available in the market that met such requirements. Table 1 summarizes the main converters features, where it can be noticed that the maximum bandwidth of the proposed SDR is restricted to 2.5 GHz.

Table 1. Summary of the main FMC converts features [16,17].

Features	FMC 126	FMC 230
Bit resolution	10	14
Max. sampling rate	5 GSPS	5.7 GSPS
Maximum signal bandwidth	2.5 GHz	2.85 GHz
Number of channels	4	2
Connector FMC type	HPC	HPC

The FMC126 is a four channel 1.25 GSPS 10-bit resolution ADC daughter board compliant to the mentioned HPC VITA 57.1 standard. Its design is based on four EV10AQ190 Quad 1.25 GSPS ADC chip-sets [18], which are characterized by Double Data Rate (DDR) Low Voltage Differential Signal (LVDS) outputs [16]. Moreover, each core can operate either individually, in pairs or even all interleaved together, which in the end enables sample rates of 1.25, 2.5, and 5 GSPS, respectively. To achieve the gigabit rates reported on this work, the ADC had its reference sample clock configured to 2.5 GHz, which can be locked by either an external clock source or an internal IC clock circuit.

The FMC230 is a two channel 14-bit DAC daughter board, which is controlled through a single Serial Peripheral Interface (SPI) communication bus [17]. Such board design is based on two AD9129 single-channel 14-bit 5.7 GSPS DAC with DDR LVDS inputs [19].

Since such boards are intended to work at gigabit rates, special attention was taken in their clock distribution. The clock distribution is carried out by a AD9517 chip from Analog Devices (AD) [20], which by default has an internal PLL and a 2500 MHz internal VCO. However, as discussed in Section 2.3, the VCO can also be provided externally.

As in FMC126, the FMC230 sample clock is provided by the mentioned AD9517, but its internal PLL reference is configured at 30.72 MHz rather than 100 MHz, which in the end achieves an internal VCO locked at 2457.60 MHz (rather than the 2500 MHz).

2.3. ADC/DAC Integration with FPGA

The integration of ADC/DAC with VC707 board required hardware fine tuning to achieve the desirable data rates. Thus, this section addresses thoroughly the steps taken to achieve such integration. It is important to mention that 4DSP already offers a VC707 reference design to serve as basis, but multiple iterations were necessary to achieve the proposed mmWave SDR testbed.

Firstly, the reference design FMC slots were changed to have both boards working simultaneously using the same FPGA. Secondly, since both boards use FPGA Ethernet as an initialization and data interface, a physical connection had to be considered onto the proposed SDR testbed. Thirdly, sampling rate adjustments were performed, because both ADC/DAC had different default values which, in the end, would lead to loss of samples. Lastly, both DAC and ADC data path were changed to be transmitted in real-time rather than using the default memory architecture, which temporarily allocates data in memories before transmission.

As mentioned above, 4DSP reference firmwares target Ethernet communication as the protocol to perform the initialization of both devices and to transmit/receive data. Consequently, such architecture demands a host PC to be always connected to the FPGA, which is impractical in a wireless communication transceiver. Therefore, Ethernet connection must be replaced by a new VHDL entity capable to play back initialization commands that were previously sent by the MAC engine to the FMC board entities (*stars*). This approach is highlighted in the block diagram illustrated in Figure 4, where all the connections in red are the ones that must be removed from the original FMC230 and FMC126 reference designs. With such changes, the MAC engine *star* is now just considered to generate clock and reset signals to the others *stars* and, due to this fact, it has been decided not to be removed.

In this work, the aforementioned new entity is a processor engine, which allows the configuration commands of both converters to be sent (in real-time) from the FPGA to either ADC or DAC. Nevertheless, the design and implementation of a processor hardware engine architecture is not a straightforward task. It is usually composed of a *microBlaze* processor, peripherals, reset, clocking and debugging blocks, which are connected together using an AXI4-Stream protocol [21]. Since the *microBlaze* is a soft-core processor with 32-bit RISC architecture and AXI interfaces, the developed entity had AXI in mind for easy, direct access to fabric and hardware acceleration [22]. It is worth noting that this interface/protocol is not a bus, and thus a custom user block was necessary to convert the AXI data coming from the processor engine to the bus format given in Figure 4.

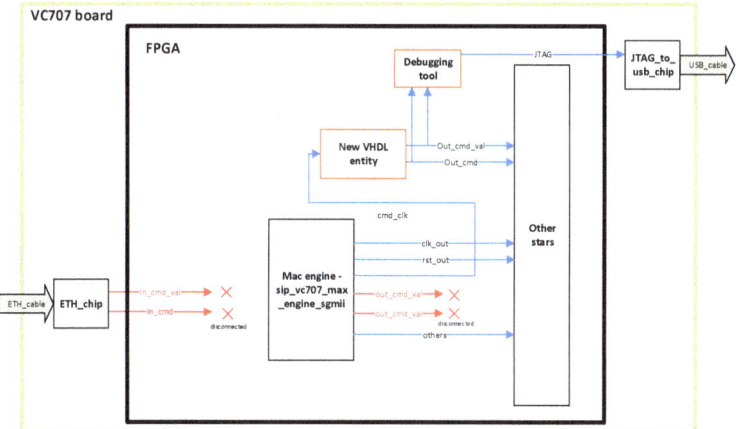

Figure 4. Block diagram of the adopted modifications common to both FMC126/FMC230.

The VC707 board has two instances of the FMC HPC VITA 57.1 mentioned in Section 2.1. However, the VC707's HPC2 connector is not fully populated, e.g., it has a reduced pin-out in comparison with the standard HPC connector. Consequently, the FMC230 placed on such connector would only have one out of its two DAC channels available. On the other hand, if FMC126 would be connected to it, only three out of its four ADC channels could be available. Considering that most relevant modern digital modulation techniques employ I/Q modulations to achieve higher spectral efficiency, two channels are required in both DAC/ADC boards. Therefore, FMC230 and FMC126 have been connected to the FMC1 and FMC2 slots of the VC707 board, respectively, as can be seen in Figure 3.

Figure 5 depicts the final firmware that shows how the configuration commands are exchanged among the top level board entity (*stars*) from both DAC/ADC boards.

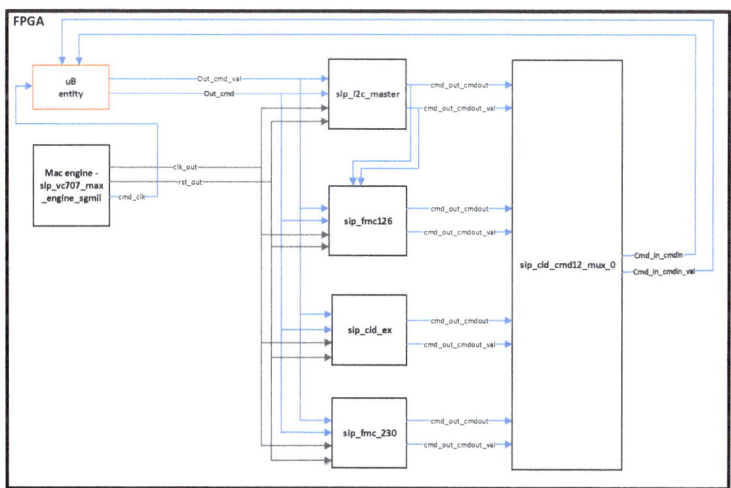

Figure 5. Block diagram of the developed transceiver firmware.

The FMC126 board includes not only the Quad ADC IC core itself, but also other peripherals, such as a clock tree IC circuit and a I2C communication for ADC control purposes and temperature monitoring [16]. Since the FMC126 reference design foresees its use in the HPC1 connector, the I2C clock (*SCL*) and data (*SDA*) signals must now be driven to the second HPC2 connector. This can be achieved by changing VC707's I2C switch, PCA9548A [23]. As depicted in Figure 6, the *sip_i2c_master* entity should route signals either to FMC1 or FMC2 connectors depending on I2C channel selection.

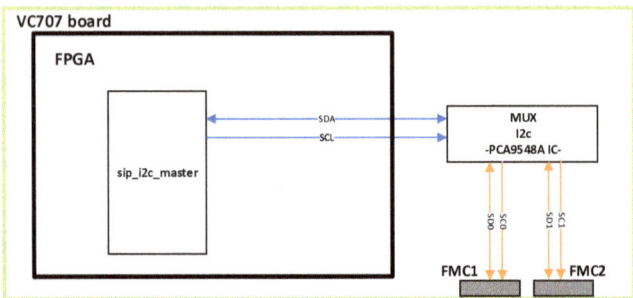

Figure 6. Illustration of the I2C communication towards VC707's FMC1 and FMC2.

2.4. ADC/DAC Sample Rate Synchronisation

With both ADC/DAC fully integrated in VC707 FPGA, considerations on their data sampling process was studied and addressed. Such process at the receiver might be subject to severe timing imperfections such as erroneous information acquisition or transmitted samples loss depending on both DAC/ADC sampling rates difference [5]. The erroneous information acquisition can occur when ADC rate is higher than DAC, whereas loss of transmitted samples might happen if ADC rate is lower than DAC. That said, sampling rate of both converter boards considered should match.

The easiest solution, found in this work, to meet the above requirement is to configure the clock tree of the FMC230 to drive a 1.25 GHz clock signal into both AD9129 cores. This would be a very straightforward task if the internal VCO of the AD9517-1 could generate a 1.25 GHz operating frequency (maximum rate achievable by FMC126 without phase interleaving) or integer multiples of this frequency. In Figure 7, it is seen that the DAC sample clock, driven by the $fDAC\_CLK$ path, is provided by the $OUT2$ output of AD9517-1. This port provides clock frequencies from 2457.6 MHz down to 12.8 MHz, which are the result of configuring either both block divisors VCO_div and $div1$ or, alternatively, bypassing VCO_div and using $div1$, or vice versa. However, only integer values are acceptable in such block dividers, and, thus, is not possible to obtain a divisor combination where $OUT2$ port is at 1.25 GHz. Therefore, meeting the ADC sample clock in the FMC230 is only possible considering an external signal as sample clock reference.

As a workaround, the clock tree of the FMC126 has been configured to enable a Clock Output (CO) signal, with Low-Voltage Positive-referenced Emitter Coupled Logic (LVPECL) waveform type, at 1.25 GHz, to be connected to the Clock Input (CI) port of FMC230. Although the AD9517-3 circuit can be easily configured to enable $OUT1$ as CO, this port is not connected to any external connector in the FMC126 board. To overcome this, an additional SSMC connector has been soldered to the CO port pad in the reverse side of the device PCB.

2.5. FMC126/FMC230 Data Interface

It is well known that operating FPGA logic at GHz clock frequencies is impractical, since very low rising interval times result into failing design timing requirements and data integrity is not ensured. Looking at current solutions, the FPGA clock performance barrier is around the 500 MHz. Hence, 4DSP adopted a parallel data path architecture in both original FMC230/FMC126 firmwares, which reduces significantly the required data clock frequencies to MHz. For example, from the FMC230 functional block diagram (see Figure 7), two parallel data DDR interface paths come from the FMC connector ($DAC0\_p0\_p/DAC0\_p0\_n$ and $DAC0\_p1\_p/DAC0\_p1\_n$—operating at $fDACCLK/4$), and are then multiplexed into a single data stream by the assembler entity, which is clocked at $fDACCLK/2$, where $fDACCLK/$ represents the DAC sampling rate value (F_s).

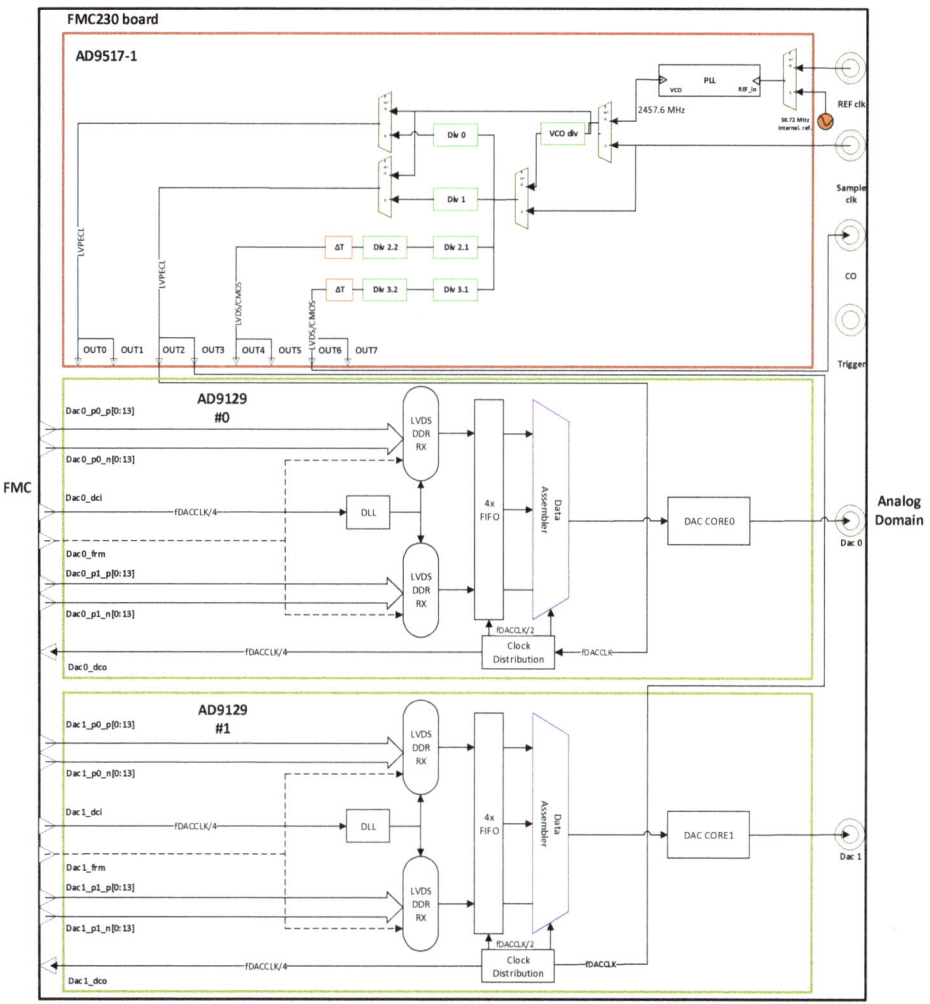

Figure 7. Detailed FMC230 functional block diagram.

A detailed block diagram of the FPGA data connectivity with DAC's firmware is depicted in Figure 8. In the figure, it is verified that data source is generated in the *Ad9129_wfm_inst0* entity, driven by the *wfm_out_data* bus signal (16 words of 16bits each) clocked at *txclkdiv8*. Such clock value is equal to $DAC0\_dco$ value divided by 4 or DAC $\frac{F_s}{16}$. Since AD9129 bit resolution is 14 bits, the two most significant bits are then discard in the *Ad9129_io_buf_v7_inst0* entity. In this entity, the 256 data bit bus (*Odata_reg*) is serialised, using two sets of fourteen parallel to serial structures of Oserdeses2 (high-speed source-synchronous output fabric interfaces) operating in DDR mode. This results in two buses of 14 bits ($DAC0\_p0\_p/DAC0\_p0$ and $DAC0\_p1\_p/DAC0\_p1$), clocked at *txclkdiv2* clock, which are then connected to the FMC230 board through the FMC1 pin-out connection (see Figure 7).

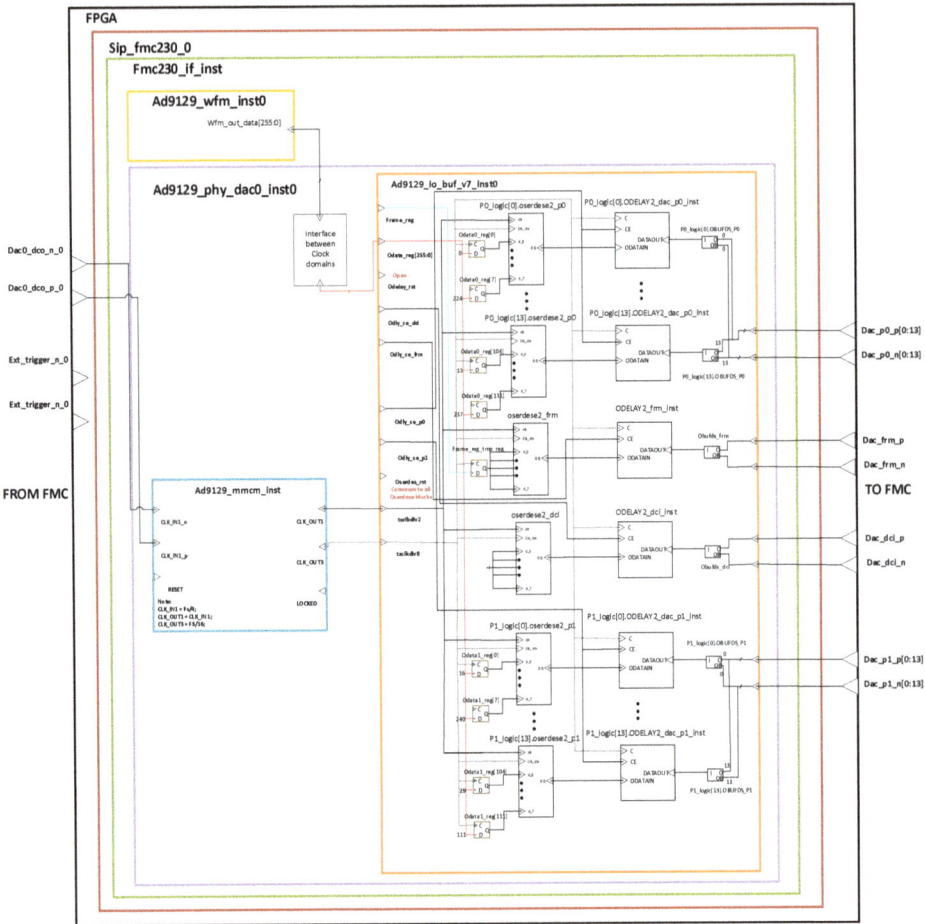

Figure 8. Block diagram of the data interface between the FPGA and FMC230 for the DAC0.

The data interface of the FMC126 with FPGA is rather similar to the FMC230 ones, but here eight parallel data paths are considered instead. That is, data go to the FMC230 on a 256 bit bus, whereas data come from the FMC126 on a 128 bit bus. This leads to a mismatch of the clock value in both DAC and ADC firmware channel paths. For example, considering a target sampling rate of 1.25 GHz, buses are clocked at 78.125 MHz and at 156.25 MHz for FMC230 and FMC126, respectively. To overcome this issue, the number of data paths in the FMC230 firmware were reduced to 8 (matching the ones from the FM126), and thus entities, namely, *ad9129_mmcm_isnt*, *ad9129_phy0_dac0_inst0*, and *ad9129_io_buf_v7_isnt0*, have been updated accordingly, where *txclkdiv8* value is now given by $\frac{F_s}{8} = Dac\_dci/2$.

The proposed communication transceiver prototype is based on the digital IF architecture shown in Figure 9. Such architecture offers more flexibility when compared to traditional RF architectures, and is not sensitive to DC offset, LO leakage and flicker noise. Moreover, since I/Q up-conversion/down-conversion is digitally performed in an IF stage, negative effects induced by IQ imbalances, critical in advanced modulation techniques, are therefore limited. Note that both Digital Up-Conversion (DUC) and Digital Down-Conversion (DDC) stages are performed to relatively low center IF frequencies (limited by the sampling rate of both DAC/ADC); however, up-conversions to mmWave can be accomplished by an external analog mixer, as discussed in Section 2.6.

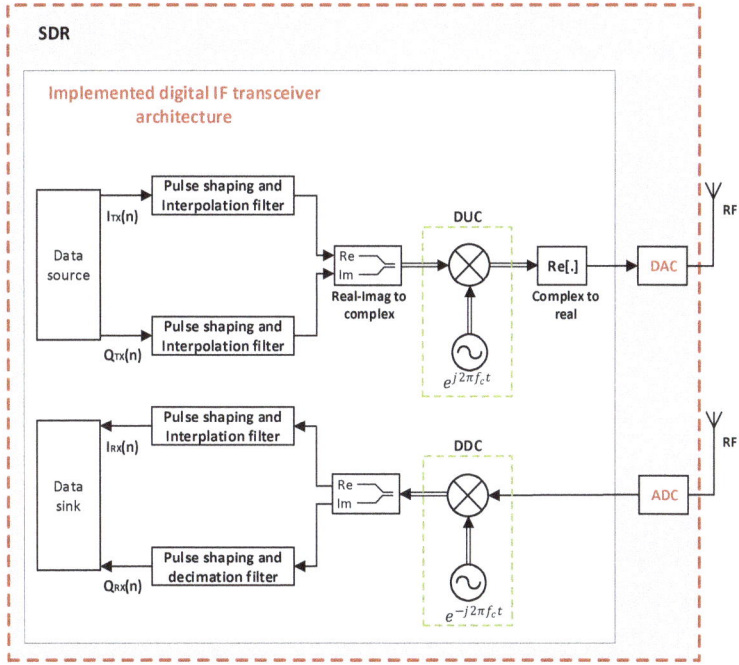

Figure 9. Block diagram of the proposed digital IF RF transceiver architecture.

As previously stated, the testbed data source is modulated with a real-time OFDM Transceiver (TRX) engine, which was developed in *Xilinx* System Generator (SysGen) environment. SysGen is the state-of-the-art tool for the design, testing, and implementation of high-performance DSP algorithms on FPGAs. It enables a rapid prototyping of very complex FPGA design by considering the *Mathworks* model-based design environment, namely *Simulink*, with total abstraction of the chip-set complexity [24,25]. In this context, OFDM TRX engine (testbed) was packed in SysGen into a customized IP to be integrated into the previously discussed transceiver firmware.

However, to design and implement the aforementioned communication system, it is required to understand how a real-time data source engine would be integrated to the data interface of both ADC/DAC reference designs (eight parallel processing paths at 156.25 MHz). In this context, at the transmitter, a DSP interpolation filter is necessary, which would be capable of interpolating the sample rate of 156 MHz to the DAC sampling rate of 1.25 GHz. Such solution was implemented using Finite Impulse Response (FIR) filter banks (clocked at 156.25 MHz) by following the polyphase decomposition algorithm reported in [26]. This solution increases the sampling rate and attenuates the generated spectral images from the up-sampling process using a low-pass filter, which also shapes the transmitted signal bandwidth [5]. On the other hand, on the RX side, the counter part must be implemented. That is, ADC sampling rate is decreased using a down sampling operation followed by a low-pass filter, removing undesired components that would otherwise overlap (aliasing) into the used band. These two operations together result in a decimation filter [5]. It is worth noting that for the I/Q up-/down- conversion 16 distributed FIR filters are necessary.

A filter bank is characterized by a parallel arrangement of low-, band-, or high-pass filters, required to decompose the spectral content of a certain signal into multiple sub-bands [27]. When the filter bank is implemented based on a low-pass prototype filter, which corresponds to the 0th band, the other uniform sub-bands are obtained by frequency shifting the prototype filter frequency response [28]. In this case, it is known as DFT filter bank and the graphical representation of its frequency response is illustrated in Figure 10.

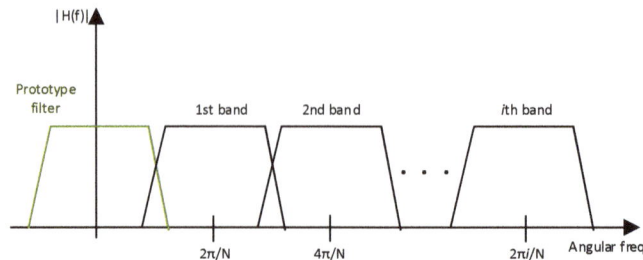

Figure 10. Graphical representation of uniform filter bank (image adapted from [28]).

Furthermore, the polyphase decomposition algorithm [26] can be employed to enable an efficient real-time implementation of uniform filter banks, while ensuring the perfect-reconstruction of the analysis/synthesis process.

Since eight parallel signal processing paths are required per I/Q path, eight analysis/synthesis FIR filters were carefully designed and implemented, considering the following procedure:

1. **Filter response type**: It is well known that pulse shaping with raised cosine is characterized by a zero-Inter-Symbol Interference (ISI) [5]. That is, despite its impulse response extending over several other symbol periods, at the decision points, there is neither constructive nor destructive interference. Therefore, in this work, a square root raised cosine matched filter pair was considered, which have the same zero-ISI properties of the raised cosine [5].
2. **Filter order and frequency specifications**: The interpolation/decimation filter was designed with the aid of the *Matlab* FDAtool algorithm. In this tool, the normalized LPF cut-off bandwidth was set to 0.125, which is the relation between the clock rates of a single processing path and the DAC core. Moreover, both *Kaiser* window Beta, and filter roll-off values were set to 2 and 0.25, respectively, since this configuration was found to be the best compromise between passband ripple, stop-band attenuation, and transition band. Finally, filter order was chosen to be 62, since it was also a compromise among system complexity, stop-band attenuation, and transition band.
3. **Distributed implementation of interpolation/decimation**: Both filter banks were implemented using multiple parallel FIR digital filters.

Additionally, both IQ up-/down- conversion stages can be performed by the multiplication of the desired signal by a complex exponential operating at a certain center frequency (f_c), as it is illustrated in Figure 9. At the transmitter, the baseband signal spectrum is frequency shifted to be centered at (f_c), resulting in a non-symmetric spectra [5]. Consequently, applying $e^{-j2\pi f_c}$ at the receiver, will move the transmitted signal spectra to be centered at DC, thus converting the received signal back to baseband.

Such complex exponential function was implemented using a Coordinate Rotational Digital Computer (CORDIC) DSP algorithm, which is responsible for implementing an IQ-mixer with angular frequency proportional to the desired f_c value. However, since both DAC/ADC converters employ an eight parallel processing path technique, similar to the FIR filter introduced above, a distributed CORDIC with eight paths operating at 156 MHz is also required to perform either an up-conversion or down-conversion. To this end, the distributed CORDIC must generate samples of the same complex exponential but at shifted sample instants in each processing path. It is worth noting that each CORDIC path in the TX FPGA firmware should be connected to the parallel structure of OSERDES, in order to perform the data serialization required in the data interface of the FMC230. In the RX firmware, to shift the signal back to baseband, a similar distributed CORDIC approach was employed but with negative phase increment. Then, it was connected to a parallel structure of ISERDES, which enables data parallelism and consequently reduction of the data clock.

Finally, to integrate the developed SysGen OFDM with the transceiver firmware, it was necessary to package and export such modulation into two distinct TX/RX IP customized blocks. That can be seen in the high-level block diagram illustrated in Figure 11.

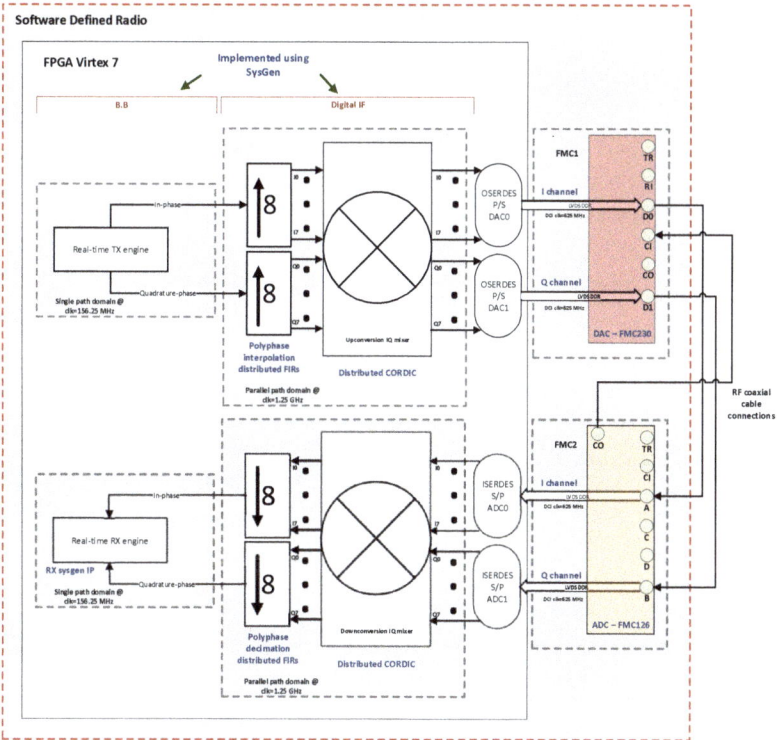

Figure 11. Block diagram of the implemented digital IF RF transceiver architecture.

2.6. RF Front-End Overview

With both ADC/DAC integration in FPGA and OFDM engine working seamlessly, this work next step was to develop the mmWave RF front-ends. The used front-end is a fully custom-made solution composed by several analog components connectorized, as illustrated in Figures 12 and 13.

On the transmitter side, it was considered a typical direct-conversion (to avoid IQ imbalance effects on transmitted signals) also known as homodyne architecture (see Figure 12). It is composed of a PLL operating at 15 GHz, a multiplier by 4, which up-converts the LO signal to 60 GHz, and an up-conversion mixer. In addition, an external 10 MHz reference clock is required for the PLL. In addition, to avoid the non-linearities introduced by the Power Amplifier (PA), the amplification stage at RF (60 GHz) has deliberatively not been included in this study. Otherwise, it would have required the study of mitigation techniques such as Digital Pre-Distortion (DPD).

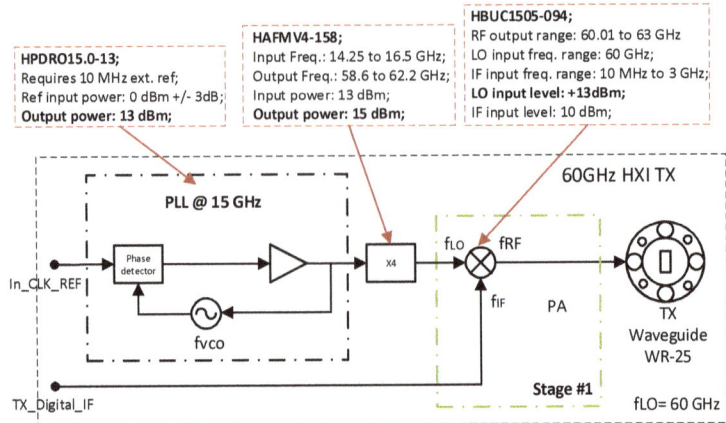

Figure 12. mmWave RF transmitter with homodyne architecture.

At the receiver, a two-stage heterodyne architecture is employed. A Low-Noise Amplifier (LNA) at RF is coupled to the antenna, followed by a down-conversion to a 6 GHz IF stage, performed by a first mixing stage process with a 54 GHz LO signal. In a second down-conversion stage, received signals are shifted to IF frequencies. Therefore, unlike the transmission architecture, two PLLs operating at 13.5 and 6 GHz are therefore required, as can be seen in Figure 13. Furthermore, the external clock references for both TX/RX PLLs are generated by two independent PRS10M rubidium oscillator. Such reference clock signals are characterized by very low phase noise and very high frequency stability (≈0.05 ppb), being, therefore, very reliable clock sources.

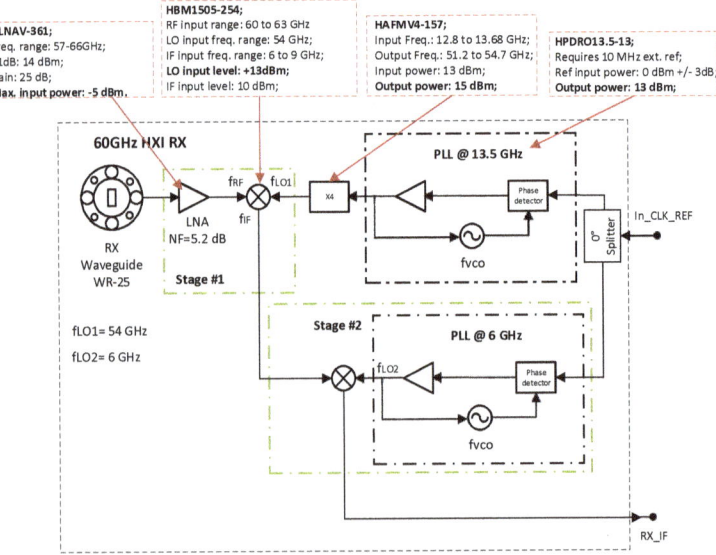

Figure 13. mmWave RF receiver with superheterodyne architecture.

3. Results Discussion

In this section, three sets of measurements are reported and analyzed. In particular, focus on the performance of ADC/DAC integration in FPGA is given. Then, performance comparison of different IF frequencies using the mentioned SDR system with OFDM engine is conducted. In the end, the proposed system (mmWave 5G testbed) performance is addressed with EVM and received scatter constellation analysis.

3.1. ADC/DAC Characterization

The FMC126 and FMC230 boards with their design modifications were characterized and validated with the usual figures of merit such as SINAD, ENOB and SFDR metrics.

With that in mind, an oscilloscope and a Spectrum Analyzer (SA) were used to measure the time and frequency responses of the DAC at multiple tones. With regards to the ADC, a signal generator was used to generate sine-wave signals with tuneable amplitude and frequency, which allowed both time and frequency responses assessment.

The DAC performance assessment was evaluated with the DAC's SFDR, which is measured as the difference between RMS power level of the transmitted tone and the most significant spur signal, presented in the DAC's frequency response. The SFDR measurement result, for the FMC230, can be seen in Figure 14 considering the D1 channel and a frequency tone of 134.3 MHz.

Additionally, the FMC230 was also characterized with SINAD, SFDR, and ENOB figures of merit for various frequency tones. Such results are summarized in Table 2 and plotted in Figure 15a,b, considering a sampling rate of 2456 MHz and DAC's D1 output channel. The average values for SFDR, SINAD, and ENOB measurements are 56.93 dB, 69.61 dB, and 11.34 bits, respectively, which according to 4DSP datasheet is within DAC performance boundaries.

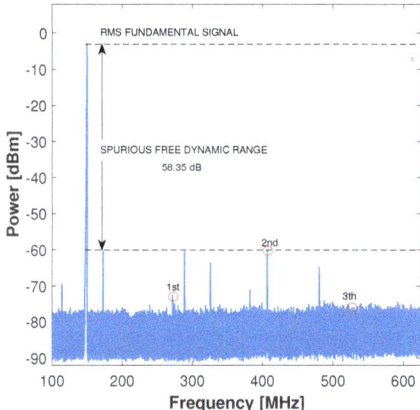

Figure 14. DAC SFDR for a tone signal of 134.3 MHz, considering the DAC's D1 channel.

Table 2. FMC230 performance boundaries versus frequency tone, for the D1 channel.

Freq. [MHz]	Carrier Power (*) [dBm]	SFDR [dB]	SINAD [dB]	ENOB [bits]
18	−3.5	54.2	75.48	12.25
134.3	−2.68	58.35	64.77	10.47
282.8	−4.37	57.36	68.75	11.13
589.3	−3.16	57.84	71.03	11.51
893.9	−4	53	68.05	11.01

(*) Average value, considering a SA resolution bandwidth of 3 kHz.

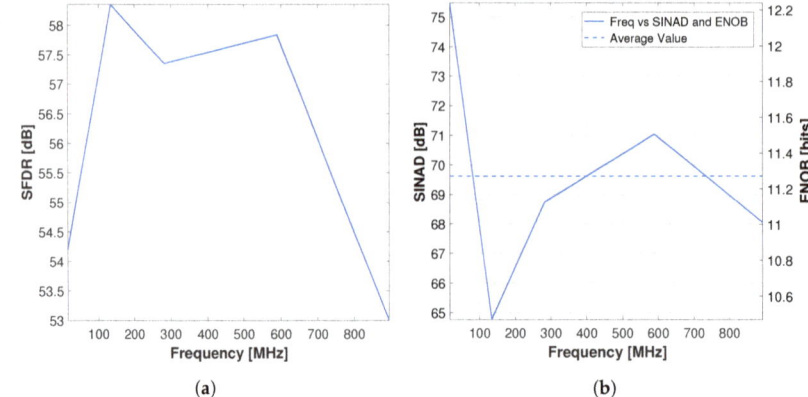

Figure 15. FMC230 performance boundaries versus frequency tone: (**a**) SFDR; and (**b**) SINAD versus ENOB.

With FMC126 in mind, a similar experimental setup was used to assert this ADC board performance. Consequently, the reference signal connected into both ADC's channel A and B, was characterized with 72.5 dB of SFDR, which is limited by the SNR of the considered signal generator equipment (Figure 16a), for a signal power level of −2 dBm (ADC full-scale input value).

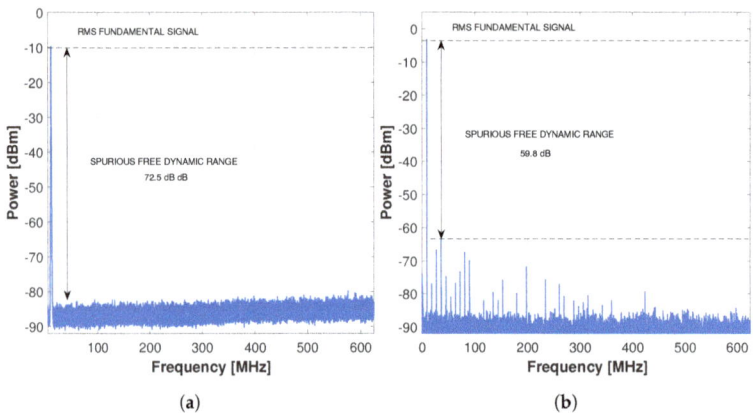

Figure 16. ADC SFDR measurement results at: (**a**) the input; and (**b**) channel A output.

Moreover, comparing Figure 16a,b, it is verified that ADC sampling and quantization processes induce a performance degradation of 12.7 dB in terms of SFDR, leading to a maximum dynamic range of 59.8 dB.

3.2. Back-to-Back Performance

Following the ADC/DAC assessment, measurements with baseband integration were conducted. It is important to mention that, following the 5G pre-trials specification presented in [1], a multi-Gigabit/s real-time OFDM TRX engine was developed [29]. Such OFDM baseband engine architecture is extensively discussed in [29] and was based on the previous work reported in [30]. Such engine enhancements led to a 1 Gbps real-time transmission rate (considering 256-QAM), which fullfills next wireless generation requirements. That said, the considered OFDM main design operating parameters, particularly the FFT and data size block values are given in Table 3.

Table 3. Main parameters considered in the design of OFDM system.

Parameter	Value
FFT size block	1024
Data block size	800
Sub-carrier bandwidth	152 kHz
Modulation	16-/64-/256-QAM
Guard time interval	256 samples
Maximum data rate	1 Gbps
Nominal signal bandwidth	122 MHz
OFDM symbol duration	8.2 µs
CP duration	1.64 µs

Table 4 outlines the main SDR system specifications accomplished by integrating both OFDM modulator/demodulator, with the digital IF architecture using the disruptive modular fully pipelined hardware architecture illustrated in Figure 11.

Table 4. Main SDR system specifications.

Feature	Value
Processing BW	1.25 GHz
Transmission BW	150 MHz
ADC/DAC sampling rate	1.25 GSPS
Modulation	OFDM with 4-/16-/64-/256-QAM
Max spectral eff.	8 bits/s/Hz/user
IF freq.	DC-612 MHz
User configuration	Single
Antenna configuration	SISO
Number of DAC/ADC channels	2/3
Resolution of DAC/ADC	14/10 bits

Firstly, the TRX performance of such SDR was evaluated in B2B configuration using EVM as QoS assessment metric. In Figure 17, an example of the transmitted OFDM signal spectrum was obtained considering a frequency centered at 312.5 MHz (IF4) and selected digital gain of 7.

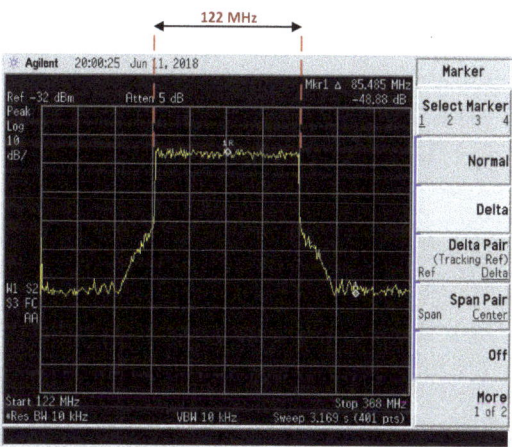

Figure 17. Frequency spectrum of the transmitted OFDM signal, for an IF4 and gain select of 7.

Additionally, numerous OFDM IF possibilities were considered, such as zero IF (DC), IF2 (156.25 MHz), IF4 (312.5 MHz), and IF6 (468.75 MHz). Figure 18 shows the output power transfer curve as a function of the selected gain and IF modulations. It was verified that OFDM modulation centered at IF2 has less signal attenuation than the remaining IF frequencies, leading to higher DAC output power.

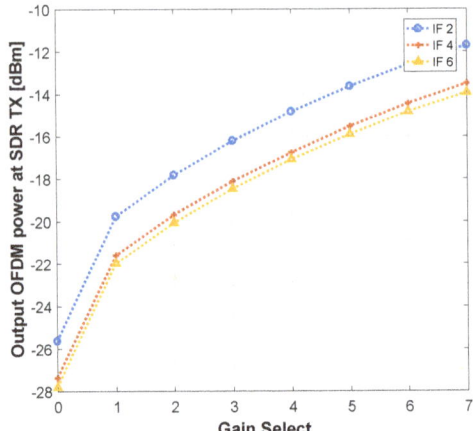

Figure 18. OFDM transmitted signal power versus digital gain and IF frequency.

In Table 5, the EVM results are summarized for the maximum digital gain of 7 using four different IF configurations. Additionally, when either I or Q channels are used to transmit and receive data, EVM results are denoted $EVM\ I$ or $EVM\ Q$, respectively. That is, connecting DAC0 to ADC0 is considered as the I branch, and connecting DAC1 to ADC1 channels is the Q branch. Additionally, for a quadrature transmission, both I/Q channel branches are used simultaneously. These results show that a minimum EVM of approximately -42 dB is obtained for all low IF modulations, except for zero-IF, in which synchronization algorithms fail to estimate the beginning of frame, due to the nulls obtained on both DAC/ADC frequency responses. As expected, for other IFs, no OFDM performance degradation is verified, even when an IQ transmission configuration is considered, indicating that very low distortion is present on received signal constellations.

Table 5. Summary of the OFDM system performance, in B2B configuration, for various IF values, DAC/ADC channel configurations, and max. digital gain.

IF Config.	EVM I		EVM Q		EVM IQ	
	[%]	[dB]	[%]	[dB]	[%]	[dB]
0	-	-	-	-	320.76	11.17
2	0.85	−41.29	0.77	−42.21	4.36*	−27.16*
4	0.84	−41.51	0.78	−42.10	0.87	−41.23
6	0.97	−41.22	0.90	−40.86	0.72	−42.75

(*) Received level power exceeds the ADC saturation point.

Finally, in Figure 19, it can be seen that the 4- and 256-QAM constellations do not present significant scattering distortion on the received symbols, by using IF2 frequency and a maximum digital gain. This subjective quality evaluation indicates that the OFDM SDR system is thus very accurate.

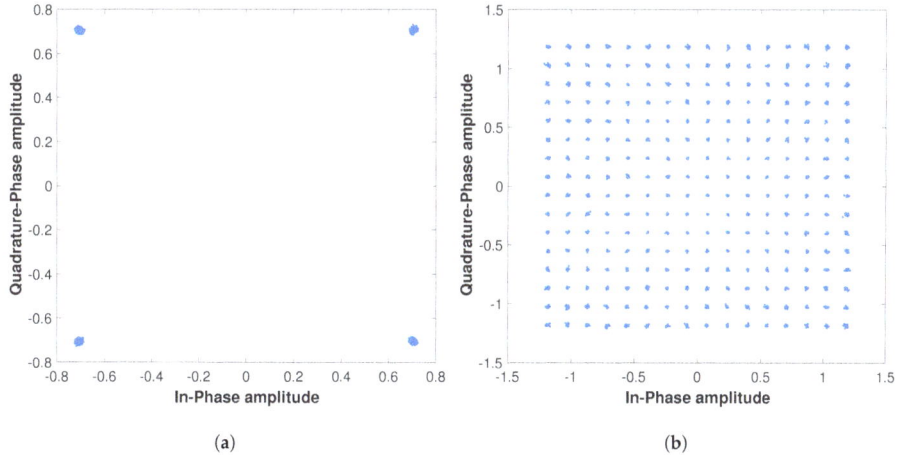

Figure 19. Received scatter constellation plots for: (**a**) 4-QAM; and (**b**) 256-QAM.

3.3. OTA 60 GHz Performance

Now, the effect of RF impairments on the quality of TRX OFDM system, outlined in Section 3.2, is addressed with 60 GHz over-the-air measurements under different conditions of RX SNIR and employed modulations. With possible OFDM performance degradation due to multipath effect in mind, the testbed was firstly set inside an anechoic chamber under LOS condition. With such controlled environment, it is possible to accurately assess the RF front-end impact on the previously discussed OFDM communication system. The system QoS was assessed through EVM figure of merit and spectral efficiency per stream analysis. In the end, testbed results with 25 dBi antennas show a maximum distance range of 74.5 cm.

A Bit Error Rate (BER) estimation was accomplished by using the relation between this metric and EVM, given in Table 6 [30]. The EVM results were computed for each OFDM IF modulated signal under different SNIR in either the presence or absence of CFO. For example, Table 7 shows the measured average EVM values for an IF OFDM modulated signal at 312.5 MHz (IF4). The minimum average EVM value, −32.99 dB, is verified for an input power of −14.83 dBm when both TX/RX devices are clocked using different sources. Such value together with Table 6 indicates that the RF front-end can handle QAM modulations up to 256-QAM, meaning that 1 Gbps of data transmissions are possible with the proposed mmWave testbed.

For quality assessment of the proposed testbed the EVM value, 4-, 16-, 64-, and 256-QAM received constellations, are depicted in Figure 20a–d, respectively. These results present a slight scattering distortion on the received symbols when compared with the B2B configuration (see constellations of Figure 19). However, even for 256-QAM, all constellations exhibit a well defined point scattering area, which indicates a low probability of decoding erroneous bits.

Table 6. Relation between EVM, BER, and digital modulation, according to Ribeiro and Gameiro [30].

BER	Modulation	EVM
10^{-3}	4-QAM	−10 dB
	16-QAM	−17 dB
	64-QAM	−24 dB
	256-QAM	−32 dB
10^{-4}	4-QAM	−12 dB
	16-QAM	−19 dB
	64-QAM	−26 dB
	256-QAM	−34 dB

Table 7. SNIR vs. EVM results using IF4 OFDM modulation for both a shared and independent clock.

Gain Sel.	TX Input Power [dBm]	Analog RX SNIR [dB]	EVM † [%]	[dB]	EVM ‡ [%]	[dB]
0	−25.63	28.70	6.83	−23.30	5.40	−25.34
1	−19.76	33.79	3.67	−28.72	3.01	−30.42
2	−17.83	36.00	3.11	−30.13	2.62	−31.60
3	−16.21	36.67	2.74	−31.21	2.30	−32.52
4	−14.83	37.81	2.35	−32.58	2.23	−32.99
5	−13.64	38.48	2.50	−32.36	2.31	−32.70
6	−12.62	38.08	2.69	−31.36	2.26	−32.87
7	−11.72	39.00	2.55	−31.87	2.37	−32.46

(†) Shared clock source configuration; (‡) Independent clock source configuration.

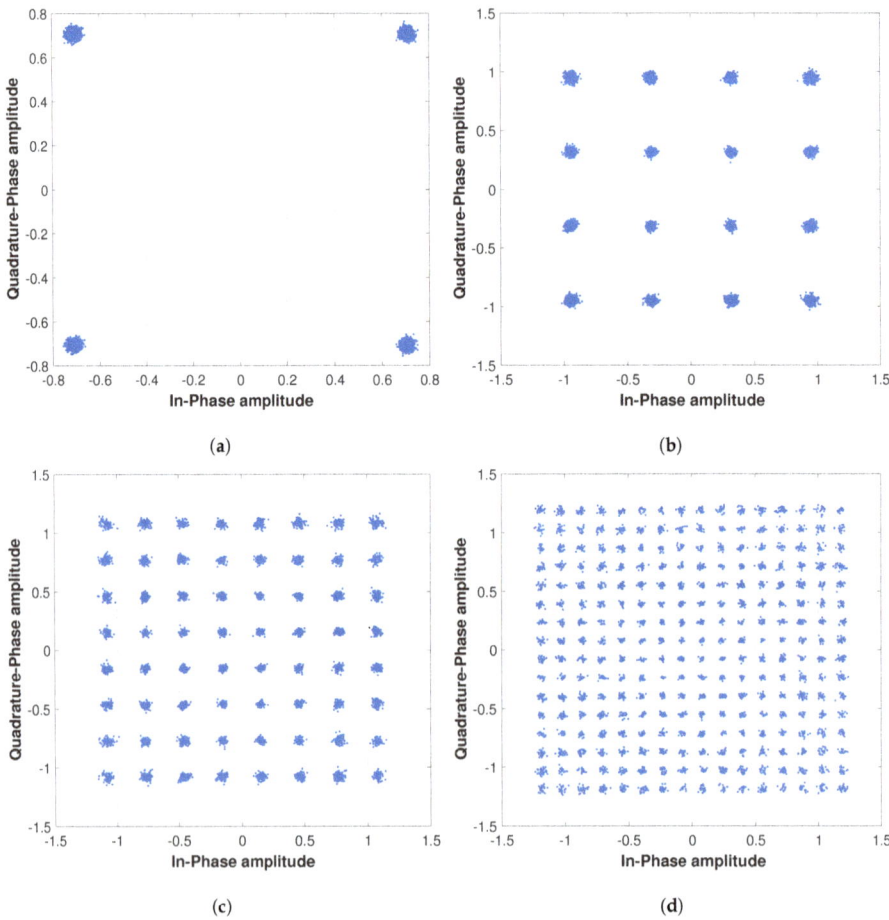

Figure 20. Received scatter constellation plots for: (a) 4-QAM; (b) 16-QAM; (c) 64-QAM; and (d) 256-QAM, considering an IF4 OFDM power transmission of −14.83 dBm on the mmWave RF front-end.

Furthermore, Table 8 shows the minimum average EVM and its degradation for the remaining IF OFDM modulations. For IF values of 312.5 and 468.75 MHz, EVM is below 3%, representing a performance degradation below 2% when compared to a back-to-back configuration. This is quite remarkable, since the EVM values at 60 GHz are lower than the 2.5% required by sub-6 GHz Wi-Fi (IEEE 802.11.ac) to employ 256-QAM OFDM transmissions [31]. To the best of authors' knowledge, this spectral efficiency per stream and SNR results go significantly beyond the state-of-the-art, when compared to the current mmWave testbeds. Finally, for the proposed link budget the received power versus digital modulation, for an BER of 10^{-3}, is presented in Table 9.

Table 8. Performance results of EVM for mmWave RF front-end.

IF	EVM %	EVM [dB]
2	5.76	−24.67
4	2.41	−32.75
6	2.90	−31.23

Table 9. Minimum received power at RX vs. digital modulation, using mmWave RF Front-End.

Modulation	Received Power [dBm]
4-QAM	−84.09
16-QAM	−66.65
64-QAM	−59.09
256-QAM	−48.35

Finally, in Figure 21, it is shown the input power operating range of the mmWave RF front-end considering IF4. It can be seen that such system operates with a relatively wide input dynamic power range considering an EVM threshold of −10 dB, which is enough for a error-free QPSK demodulation.

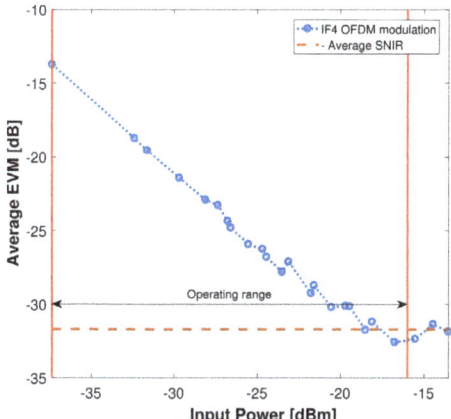

Figure 21. mmWave RF front-end input power versus EVM, for IF4 OFDM with absence of CFO.

From the above results, it is demonstrated that it is possible to successfully use the mmWave spectrum to achieve multi-Gigabit/s, while considering relatively high spectral efficient modulations, and signal bandwidths greater than 100 MHz. For example, in literature only the 28 GHz testbeds reported in [32–35], consider a modulation bandwidth higher than 100 MHz using real-time baseband processing, and considering the 5G NR waveform as the transmission scheme (OFDM). On the other hand, at 60 GHz, it is evident a lack of testbeds that comply the minimum technical performance requirements of IMT 2020 for 5G. Considering the proposed OFDM testbed, the 5G peak spectral efficiency value of 7.8 bit/s/Hz [1] was achieved, and thus, the gap verified in testbeds operating at 60 GHz has been fulfilled. In other words, no other prototype system can process a signal BW of 150 MHz in real-time with modulation orders up to 256 QAM, using OFDM as transmission scheme in an over-the-air scenario. It is worth noting that even the listed 28 GHz testbed systems, do not meet the DL specification of 7.8 bit/s/Hz.

4. Showcasing: From GbE-Based to UHD Multi-Stream Video

In order to make the testbed appealing to the general public, a video base demonstrator was put together, as it is depicted Figure 22. Such showcase was a joint-collaboration with the Multimedia group in IT Leiria that developed a 180° field of view system as in [36].

The UHD multi-stream showcase has been presented at several key forums, creating great impact with both scientific community and industrialist. Such video content was generated independently with the 180° field of view camera setup [36], which is composed of 9 UHD cameras responsible for 20° video acquisition each. Due to the agglomerate high data throughput, a GigabitEthernet connection is required to route all 9 UHD cameras data from the encoded stream outputted from the Raspberry Pi to the transmitting FPGA. That said, a GbE switch merges all 9 Raspberry Pi ethernet connections into the single GbE available at TX baseband unit. At RX side, a high performance laptop with GbE interface runs VLC for real-time decoding and video displaying, the decoded video content from the RX baseband unit.

This has proven to be eye-catching and intuitive, since the visitors could directly interact with the system, e.g., by blocking the direct radio path and observe the immediate impact in both received signal constellations and received video quality. Such user interaction coupled to the fact that 9 UHD video streams were being displayed in real-time on the computer, provided the required validation of a real-time over-the-air user experience. On the other hand, for those with a technical background, there is a Graphical User Interface (GUI) displaying in real-time the received signal constellation, as well as the estimated BER values.

(a) (b)

Figure 22. Photograph of the proposed testbed: (**a**) Techdays 2017, Aveiro; and (**b**) Ciência 2018, Lisbon.

5. Conclusions

In this article, a novel software-defined radio solution for future wireless communication systems is presented and validated as a complete multi-Gigabit/s radio in-the-loop OFDM transmission scheme with mmWave 60GHz RF front-end to cope with UHD multi-streams.

The SDR hardware choice was based on the overall criteria of low time-to-market and reduced development costs. On the one hand, VC707 FPGA was the choice for the embedded DSP engine, since at that time it was the best compromise among available logic resources, highly-flexible, high-speed serial bus I/O interface, and cost. In addition, both FMC230/FMC126 were selected based on their inherent high sampling rates, bit resolution, and their compatibility with the VC707 FMC I/O interface connectivity.

Moreover, firmware connectivity issues between the embedded system and both FMC230/FMC126 have been outlined in this work, as well as the considered workarounds to enable a GSPS real-time communication system. Currently, the SDR is configured to 1.25 GSPS, which, as discussed, is 5G future-proof. In addition, both ADC/DAC converters were deeply characterized in terms of SNR, ENOB, SINAD, and SFDR, and their advantages of a digital IF RF architecture has been extensively discussed.

Additionally, a complete multi-Gigabit/s radio in-the-loop OFDM communication system was implemented for high data rate applications considering 4-, 16-, 64-, and 256-QAM, which was validated. EVM results in both scenarios were below 2%, which is quite remarkable considering the current norm for employed 256-QAM OFDM transmissions is 2.5% [31].

A complete overview on mmWave front-end was done, which together with the baseband unit form a robust communication system with rates up to 1 Gbps. The chosen use case for this work was UHD video stream that was achieved with a 180° field of view setup available at IT.

In summary, this work details the required steps to achieve a real-time testbed that meets 5G requirements on the mmWave frequency range with 1 Gbps and low latency main characteristics.

Author Contributions: Conceptualization, methodology, software, validation, formal analysis, investigation, resources, data curation, writing—original draft preparation, writing—review and editing and visualization—L.D., R.G., C.R. and R.F.S.C.; Supervision, project administration and funding acquisition—C.R. and R.F.S.C.

Funding: This research was partially supported by the Portuguese Government under the project RETIOT—Reflectometry Technologies to Enhance the Future Internet of Things and Cyber-Physical Systems, reference POCI-01-0145-FEDER-016432 (SAICT-45- 2015-03/16432), funded by FEDER within COMPETE 2020; Instituto de Telecomunicações under the projects 5G TESTBED and 5G MIMO TESTBED (UID/EEA/50008/2019).

Conflicts of Interest: The authors declare no conflict of interest.

References

1. Shafi, M.; Molisch, A.F.; Smith, P.J.; Haustein, T.; Zhu, P.; Silva, P.D.; Tufvesson, F.; Benjebbour, A.; Wunder, G. 5G:A Tutorial Overview of Standards, Trials, Challenges, Deployment, and Practice. *IEEE J. Sel. Areas Commun.* **2017**, *35*, 1201–1221. [CrossRef]
2. Quadri, A.; Zeng, H.; Hou, Y.T. A Real-Time mmWave Communication Testbed with Phase Noise Cancellation. In Proceedings of the IEEE INFOCOM 2019—IEEE Conference on Computer Communications Workshops (INFOCOM WKSHPS), Paris, France, 29 April–2 May 2019; pp. 455–460. [CrossRef]
3. Zetterberg, P.; Fardi, R. Open Source SDR Frontend and Measurements for 60-GHz Wireless Experimentation. *IEEE Access* **2015**, *3*, 445–456. [CrossRef]
4. Saha, S.K.; Ghasempour, Y.; Haider, M.K.; Siddiqui, T.; Melo, P.D.; Somanchi, N.; Zakrajsek, L.; Singh, A.; Shyamsunder, R.; Torres, O.; et al. X60: A Programmable Testbed for Wideband 60 GHz WLANs with Phased Arrays. *Comput. Commun.* **2019**, *133*, 77–88. [CrossRef]
5. Stewart, R.W.; Barlee, K.W.; Atkinson, D.S.W.; Crockett, L.H. *Software Defined Radio Using MATLAB & Simulink and the RTL-SDR*; Strathclyde Academic Media: Glasgow, UK, 2015.
6. Bronckers, S.; Roc'h, A.; Smolders, B. Wireless Receiver Architectures Towards 5G: Where Are We? *IEEE Circuits Syst. Mag.* **2017**, *17*, 6–16. [CrossRef]
7. Software-Defined Radio. Available online: https://www.winradio.com/home/facts.htm (accessed on 18 October 2017).
8. Kuon, I.; Rose, J. Measuring the Gap Between FPGAs and ASICs. *IEEE Trans. Comput. Des. Integr. Circuits Syst.* **2007**, *26*, 203–215. [CrossRef]
9. Field Programmable Gate Array (FPGA). Available online: https://www.xilinx.com/products/silicon-devices/fpga/what-is-an-fpga.html (accessed on 8 March 2017).
10. Malachy Devlin. *VITA 57 (FMC) opens the I/O pipe to FPGAs*; Forasach : Glasgow, UK, 2012.
11. Paper, W. *I/O Design Flexibility with the FPGA Megazzine Card (FMC)*; Technical Report; Xilinx : San Jose, CA, USA, 2009.
12. Xilinx Ultrascale. Available online: https://www.xilinx.com/support/documentation/selection-guides/ultrascale-plus-fpga-product-selection-guide.pdf (accessed on 6 November 2019).

13. Xilinx VC707. Available online: https://www.xilinx.com/products/boards-and-kits/ek-v7-vc707-g.html (accessed on 6 November 2019).
14. 4DSP FMC126. Available online: http://www.4dsp.com/{FMC}126.php (accessed on 6 November 2019).
15. 4DSP FMC230. Available online: http://www.4dsp.com/{FMC}230.php (accessed on 6 November 2019).
16. UM009 FMC12x. *FMC122/FMC125/FMC126: User Manual*; Technical Report; 4DSP : Austin, TX, USA, 2014.
17. UM023 FMC230. *FMC230: User Manual*; Technical Report; 4DSP : Austin, TX, USA, 2015.
18. EV10AQ190 Datasheet. Available online: https://www.teledyne-e2v.com/resources/account/download-datasheet/1735 (accessed on 8 January 2017).
19. AD9129. *11-/14-Bit, 5.7 GSPS, RF Digital-to-Analog Converter*; Data Sheet; Analog Devices: Norwood, MA, USA, 2017.
20. AD9517. *12-Output Clock Generator with Integrated 2.8 GHz VCO*; Data Sheet; Analog Devices: Norwood, MA, USA, 2013.
21. UG940. *Vivado Design Suite Tutorial: Embedded Processor Hardware Design*; Technical Report; Xilinx: San Jose, CA, USA, 2017.
22. UG761. *AXI Reference Guide*; Technical Report; Xilinx: San Jose, CA, USA, 2011.
23. PCA9548A. *Low Voltage 8-Channel I2C Switch with Reset*; Technical Report; Texas Instruments: Dallas, TX, USA, 2015.
24. Mittal, S.; Gupta, S.; Dasgupta, S. System Generator: The State-of-art FPGA Design tool for DSP Applications. In Proceedings of Third International Innovative Conference On Embedded Systems, Mobile Communication And Computing (ICEMC2 2008), Pesimsr, India, 11–14 August 2008.
25. Paper, W. *Model-Based Design with Simulink, and Xilinx System Generator for DSP*; Technical Report; MathWorks: Natick, MA, USA, 2012.
26. Vaidyanathan, P.P. Multirate digital filters, filter banks, polyphase networks, and applications: A tutorial. *Proc. IEEE* **1990**, *78*, 56–93. [CrossRef]
27. Mertins, A.; Mertins, D.A. *Signal Analysis: Wavelets, Filter Banks, Time-Frequency Transforms and Applications*; John Wiley & Sons, Inc.: New York, NY, USA, 1999; Chapter 6.
28. Farhang-Boroujeny, B. Filter Bank Spectrum Sensing for Cognitive Radios. *IEEE Trans. Signal Process.* **2008**, *56*, 1801–1811. [CrossRef]
29. Ribeiro, R.C.; Gomes, L.D.A.H.; Caldeirinha, R.F.S. Multi-Gigabit/s OFDM real-time based transceiver engine aiming at emerging 5G MIMO applications. *Phys. Commun.* **2019**, in press. [CrossRef]
30. Ribeiro, C.; Gameiro, A. A Software-defined Radio FPGA Implementation of OFDM-based PHY Transceiver for 5G. *Analog. Integr. Circuits Signal Process.* **2017**, *91*, 343–351. [CrossRef]
31. Verma, L.; Fakharzadeh, M.; Choi, S. WiFi on Steroids: 802.11ac and 802.11AD. *IEEE Wirel. Commun.* **2013**, *20*, 30–35. [CrossRef]
32. Sung, M.; Cho, S.; Kim, J.; Lee, J.K.; Lee, J.H.; Chung, H.S. Demonstration of IFoF-Based Mobile Fronthaul in 5G Prototype With 28-GHz Millimeter wave. *J. Lightwave Technol.* **2018**, *36*, 601–609. [CrossRef]
33. Obara, T.; Okuyama, T.; Aoki, Y.; Suyama, S.; Lee, J.; Okumura, Y. Indoor and outdoor experimental trials in 28-GHz band for 5G wireless communication systems. In Proceedings of the 2015 IEEE 26th Annual International Symposium on Personal, Indoor, and Mobile Radio Communications (PIMRC), Hong Kong, China, 30 August–2 September 2015; pp. 846–850. [CrossRef]
34. Mashino, J.; Satoh, K.; Suyama, S.; Inoue, Y.; Okumura, Y. 5G Experimental Trial of 28 GHz Band Super Wideband Transmission Using Beam Tracking in Super High Mobility Environment. In Proceedings of the 2017 IEEE 85th Vehicular Technology Conference (VTC Spring), Sydney, NSW, Australia, 4–7 June 2017; pp. 1–5. [CrossRef]

35. Final Radio Interface Concepts and Evaluations for Mm-Wave Mobile Communications. Deliverable D4.2; Project Millimetre-Wave Based Mobile Radio Access Network for Fifth Generation Integrated Communications (mmMAGIC). Availiable online: https://www.google.com.hk/url?sa=t&rct=j&q=&esrc=s&source=web&cd=1&ved=2ahUKEwiI07j-85_mAhURyIsBHQbmCC8QFjAAegQIAxAC&url=https%3A%2F%2Fbscw.5g-mmmagic.eu%2Fpub%2Fbscw.cgi%2Fd214055%2FmmMAGIC_D4.2.pdf&usg=AOvVaw3ly2g-tqAOmzKiKjsl9UcY (accessed on 30 June 2017).
36. Duarte, J.; Assuncao, P.A.A. Low-complexity acquisition of 180-degree panoramic video using early data reduction. In Proceedings of the IEEE EUROCON 2019—18th International Conference on Smart Technologies, Novi Sad, Serbia, 1–4 July 2019; pp. 1–6. [CrossRef]

© 2019 by the authors. Licensee MDPI, Basel, Switzerland. This article is an open access article distributed under the terms and conditions of the Creative Commons Attribution (CC BY) license (http://creativecommons.org/licenses/by/4.0/).

MDPI
St. Alban-Anlage 66
4052 Basel
Switzerland
Tel. +41 61 683 77 34
Fax +41 61 302 89 18
www.mdpi.com

Electronics Editorial Office
E-mail: electronics@mdpi.com
www.mdpi.com/journal/electronics

www.ingramcontent.com/pod-product-compliance
Lightning Source LLC
LaVergne TN
LVHW071948080526
838202LV00064B/6709